Bees

Shown to the Children

Ellison Hawks

Alpha Editions

This edition published in 2024

ISBN : 9789367249406

Design and Setting By
Alpha Editions
www.alphaedis.com
Email - info@alphaedis.com

Contents

ABOUT THIS BOOK...- 1 -

CHAPTER I ABOUT THE BEE...........................- 2 -

CHAPTER II THE QUEEN BEE..........................- 4 -

CHAPTER III THE DRONE...............................- 6 -

CHAPTER IV THE WORKER BEE ....- 8 -

CHAPTER V THE MICROSCOPE ....- 9 -

CHAPTER VI THE HEAD ....- 11 -

CHAPTER VII THE WONDERFUL
ANTENNÆ...- 13 -

CHAPTER VIII THE EYES...............................- 17 -

CHAPTER IX THE TONGUE AND
MOUTH PARTS ....- 20 -

CHAPTER X THE JAWS ....- 22 -

CHAPTER XI THE THORAX............................- 24 -

CHAPTER XII THE LEGS ....- 25 -

CHAPTER XIII THE FIRST PAIR OF
LEGS...- 27 -

CHAPTER XIV THE SECOND AND THIRD PAIR OF LEGS ....- 29 -

CHAPTER XV THE WINGS...- 31 -

CHAPTER XVI THE ABDOMEN ....- 34 -

CHAPTER XVII THE BREATHING APPARATUS ....- 36 -

CHAPTER XVIII THE STING ....- 38 -

CHAPTER XIX THE ANCIENTS AND BEES..- 40 -

CHAPTER XX THE HIVE ....- 44 -

CHAPTER XXI A VISIT TO A HIVE ....- 47 -

CHAPTER XXII THE CITY GATE ....- 49 -

CHAPTER XXIII THE GUARD BEES..............................- 52 -

CHAPTER XXIV WORKERS IN THE CITY ....- 54 -

CHAPTER XXV THE COMB BUILDERS..- 56 -

CHAPTER XXVI THE LIFE OF THE BEE ....- 60 -

CHAPTER XXVII THE STORY OF THE QUEEN ....- 63 -

CHAPTER XXVIII THE POLLEN GATHERERS ....- 65 -

CHAPTER XXIX THE VARNISH MAKERS ....- 68 -

CHAPTER XXX THE NECTAR GATHERERS ....- 70 -

CHAPTER XXXI THE WINTER SLEEP ....- 72 -

CHAPTER XXXII THE SWARM ....- 74 -

CHAPTER XXXIII TAKING THE SWARM ....- 77 -

CHAPTER XXXIV THE OLD HIVE AFTER A SWARM ....- 79 -

CHAPTER XXXV THE MASSACRE OF THE DRONES ....- 81 -

CHAPTER XXXVI HONEY ....- 83 -

CHAPTER XXXVII MODERN BEE-KEEPING ....- 85 -

CHAPTER XXXVIII THE BEES' ENEMIES ....- 88 -

CHAPTER XXXIX POWERS OF COMMUNICATION...........................- 90 -

CHAPTER XL BEE FLOWERS...........................- 92 -

CHAPTER XLI POLLEN ....- 95 -

CHAPTER XLII BEES AND FLOWERS ....- 97 -

CHAPTER XLIII HOW FLOWERS PROTECT THEIR NECTAR...............................- 99 -

CHAPTER XLIV HOW FLOWERS ARE
FERTILISED ....- 100 -

CHAPTER XLV CONCLUSION ....- 103 -

ABOUT THIS BOOK

DEAR ANNIE AND KATIE,—When I was a little boy I often wished that my soldiers would come to life. I used to think how grand it would be if only I could have a city of little people on the dining-room table. Of course my dreams never came true, even though one day I had a brilliant idea, and wrapped a whole regiment of soldiers in flannel and put them in the oven, hoping that in this way I should find them really alive next morning!

But nowadays I have a wonderful city of tiny workers, that can be put on a table. In it there are soldiers, food gatherers, bread-makers, undertakers, and a host of others. It is ruled over by a queen, and each day the gates of the city are crowded with the workers, who pass in and out in hundreds.

Have you guessed that my wonderful city is really a bee-hive? Although I cannot command my little friends to do this thing or that, to come here or go there, yet I am quite content to leave them to their own ways, and just to watch them in their daily life, and to study their customs and laws.

In this little book I intend to tell you something about my bees. I hope that you will be interested to read what I have written, and then perhaps, later on, when you grow up, you may keep bees, and you will be able to study their wonderful ways for yourselves.

I am sure you will join me in giving our best thanks to my friends who have so kindly helped me in the preparation of some of the pictures: to Mr. W. Barker, Mr. D. Ingham, Mr. H. Mackie, Mr. G. W. Stephenson; and to Mr. J. Lambert for permission to use Plates Nos. XIV., XV., XXVII., XXIX. and XXX.

My thanks are due also to Mr. W. H. McCormick for his kindness in reading over the proofs.

Yours truly,

ELLISON HAWKS.

10 GRANGE TERRACE,
LEEDS, 1912.

CHAPTER I
ABOUT THE BEE

NO matter how small an insect may be, it is sure to teach us something interesting if we study its habits, and try to find out how the various parts of its body are used. Perhaps of all the thousands of different insects upon the earth, the most wonderful of all are Bees. When we speak of bees we generally think of those which live in the white hives we sometimes see in gardens; these are the bees kept by a man to make honey for him. You will perhaps be surprised, therefore, to learn that there are over 2000 different kinds of bees known at the present time, and that over 200 of these species are found in Great Britain. These include the different kinds of hive bees and also the wild bees, for there are races of bees just as there are races of mankind. In this little book I hope to tell you about the hive bee, or, as it is called by its Latin name, *Apis mellifica* ("the honey bee"). In the first few chapters we shall learn something about the body of the bee, and its different limbs and organs. Later on we shall consider the construction of the hive, and the habits of the bees which dwell therein.

The word insect comes from the Latin, and means "divided into parts." If you look at the body of a bee, or of any other insect, you will find that it is divided into three parts. These three divisions are respectively known as the Head, the Thorax, and the Abdomen. The head carries the *antennæ* or feelers, as they may be called; the thorax, or chest, has the wings and legs joined to it; whilst the abdomen, or hindermost part of the body, contains the stomach and internal organs.

There are three kinds of bees in a hive—the Queen, the Drone, and the Worker, and a picture of these is seen in Plate I. Only one queen bee is found in each hive, though there may be several hundred drones and perhaps 50,000 or 60,000 workers. The number of the workers and drones varies according to the size of the hive and the time of the year.

The races of bees are many, but the best known is the British bee, sometimes called the Black Bee. Why it should be called "black" no one seems to know, for, as a matter of fact, it is of a beautiful rich brown colour. Then there is the Ligurian bee, which is of a lighter shade, and has three golden bands around its abdomen, by which you will easily recognise it. The Carniolian bees are natives of Carniolia in Austria, and they also have rings, but of a lighter yellow colour, while the bee itself is not such a dark brown as the Ligurian. Carniolian bees are supposed to be very sweet-tempered, and are therefore sometimes called "the lady's bees." Whether they really are better-tempered than other races is a question, for the temper of the little insects depends a great deal upon circumstances. For instance, if spiders have been trying to get into the hive, the bees are often very cross, and it is dangerous to go anywhere near them. But should there be no trouble of this kind to worry them, the hive may be opened and the bees handled without fear.

PLATE II

From a photograph by] [E. Hawks
Queen

CHAPTER II
THE QUEEN BEE

LET us now look at Plate II., where an illustration of a queen bee is to be seen. It will be noticed that her abdomen is much longer than that of the worker or of the drone. Her head and thorax are about the same size as those of the others, but her legs are slightly longer and differently shaped.

This then is the queen of the hive, and she has, as we have seen, many thousands of subjects. We might imagine that, such being the case, she would lead a life of pleasure and enjoyment; but this is not so. In fact she is wrongly named the queen, for she does not rule over the other bees in the way we are accustomed to think of a king or queen doing. She would be better called the mother of the hive, for she is the parent of all the other bees. She never leaves the bee-city, except perhaps on one or two state occasions, so that she spends the greater part of her life in the darkness of the hive. She is waited upon and fed by her royal attendants, who also clean her and guide her over the combs. Perhaps, some time or other, if you have the opportunity of doing so, you may see the queen of some friend's hive. You will see her on the comb, no doubt, and you will notice a circle of six or more bees around her. These are her attendants, who face her and do not turn their backs to her if it can be avoided. In Plate III. is shown the queen surrounded by her attendants. They are within the circle which has been drawn on the photograph, and the arrow points to the queen. Great care is taken of the queen, for on her depends the future of the race, and so she is closely guarded as well as being tended and fed. Every one of the little workers would willingly lay down her life for the sake of the queen, were this necessary.

PLATE III

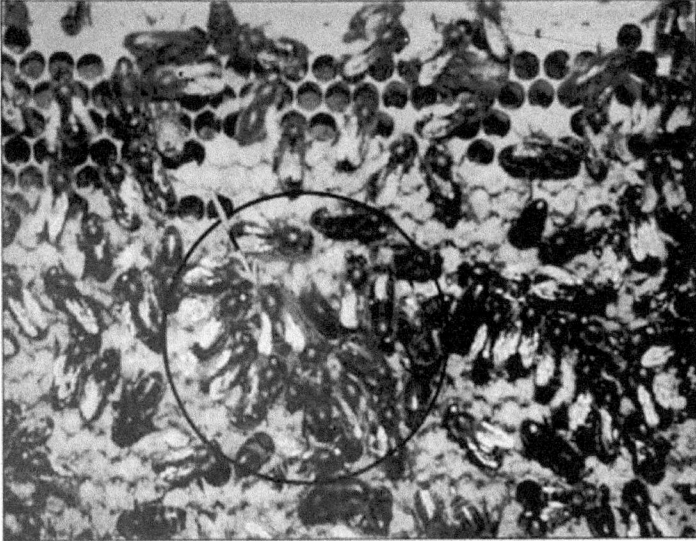

From a photograph by] [E. Hawks

The Queen Bee surrounded by her Attendants

CHAPTER III
THE DRONE

NOW a few words about the drone, or male bee, and a picture of him is shown in Plate IV. He is not so big as the queen, though he possesses a more burly appearance. Unlike the queen or the worker bees, the drone has no sting, and so you may let him crawl over your hand without fear of being hurt, even though he should become angry.

The life of the drone is a life of luxury and ease, for he does not work in the hive, neither does he gather any nectar or pollen. He is fed by the workers, and he also takes good care to help himself from the storehouses, whenever he thinks he would like a little more food. He generally finds some snug corner in the hive, away from the bustle of the city, and there sleeps till perhaps mid-day. Then at this hour, after a good meal, he sallies forth, pushes his way through the crowd of workers, and with a loud, droning noise flies away to some far-off flower, perchance, and there basks in the sun. Before the afternoon wanes, he returns to the bee-city, has another meal, and then sleeps until next day. A very lazy life is this, you will say, and I agree with you. But this life, like all good things, comes to an end, and little though the drones know it, before the winter comes they will be killed by executioners duly appointed by the other bees, and so their life of luxury will be cut short.

In appearance the drones are very beautiful, and if we watch the door of a hive, some summer day, we may see them come out to take their daily outing. Their eyes are like enormous black pearls on each side of their head, while the silky antennæ look like beautiful plumes. Their thoraxes are covered with many golden hairs, which make them look as though they were clothed in the finest yellow velvet.

As they leave the hive, they create quite a stir amongst the other bees. They care not for the sentries, and rushing out, overturn the foragers who are coming in from the fields. No notice is taken of their rudeness, but the workers go on with their various duties, no doubt thinking that ere long the day of execution will come, and that then they will be avenged.

PLATE V

From a photograph by] [E. Hawks

Worker

CHAPTER IV
THE WORKER BEE

ON Plate V. is shown an illustration of the worker bee, which is the smallest inhabitant of the hive, but, nevertheless, does the greatest amount of work. We have already seen that there are many thousands of workers in a hive, and that each one has certain duties assigned to her. All are busy, and they work as though the future of the whole hive depended on their labours—as indeed it does. Each worker seems to be trying to outdo the others, in the endeavour to see how much work she can crowd into her little life. Laziness is unknown, and should a bee become so badly injured from any cause as not to be able to work, she is put to death, for the government of the bee-city has plenty of mouths to fill, without any useless ones. To us this may seem cruel, but we must admit that it is economical.

The duties of the workers are numerous. There are the water carriers, to supply the hive with water; the nurse bees, to look after the young ones; the foragers, who gather nectar and pollen. Then there are the builders, architects, undertakers, scavengers, chemists, and soldiers. Lastly there are the house bees and the ventilating bees.

Each bee is allocated to one or another of these trades, and each one seems to know exactly how to do the work, and always seems to be doing it! There is no quarrelling as to who shall gather pollen, or who shall guard the city, for all is arranged by some mysterious law.

CHAPTER V
THE MICROSCOPE

BY the aid of a wonderful instrument called the Microscope we are able to learn a good deal about the construction of the different parts of the bee's body.

A microscope, as perhaps you already know, is a sort of strong magnifying glass, being something like a telescope, but on a smaller scale altogether. You may see an illustration of a microscope in (*a*) Plate VI. The tube of the microscope is generally about six or eight inches in length, made of metal and holding two sets of lenses. The one through which we look is at the top of the tube, and is called the eyepiece. The lens at the bottom is called the objective, for it is the lens that is nearest to the object that is to be examined. If you have a microscope of your own, or know any one who has one, you will be able to see for yourselves many of the things about which I am going to tell you. For the sake of convenience the parts of insects to be examined in the microscope are generally mounted on little slips of glass, and if you place a dead bee on a piece of glass, you will find that it is more easily handled in this manner. Some of you, however, may not have this opportunity, and so I have photographed several different parts of the bee, by the aid of the microscope, so that you will be able to understand what you will read about them.

PLATE VI

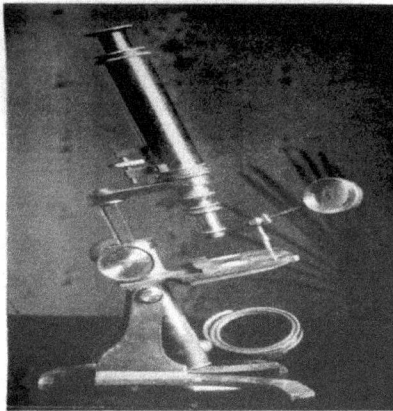

(*a*)
From a photograph by] [E. Hawks
A Microscope

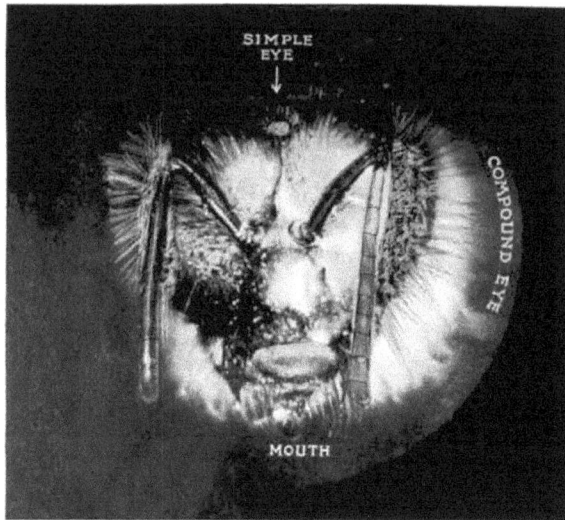

(*b*)
From a photo-micrograph by] [E. Hawks

Head of Bee

Just as the telescope has taught its users a great deal about the stars, which otherwise could not have been known, so too has the microscope shown us wonders such as we never before thought existed.

Before we consider the habits of the bees, it will be well for us to examine, and to understand, the various limbs and parts of their bodies, in order that we may the more easily trace out the manner in which the little workers accomplish their tasks.

CHAPTER VI
THE HEAD

JUST as the head of an animal is the most important part of its body, so too is it in the case of an insect.

A bee's head, as seen with the aid of a microscope, looks very peculiar, but nevertheless it is exceedingly interesting. A photograph of it is shown (*b*) on Plate VI. The head is something like a split pea in shape, with the rounded part turned to the front; it is joined to the thorax by a thin neck.

The bee has five eyes, two compound and three simple. The compound eyes are placed one on each side of the head, like the eyes of the house-fly, and the simple eyes are to be found on the top of the head. In Plate VII. the position of the eyes is shown, but only one of the simple eyes is to be seen. In addition to the eyes, the head carries the antennæ, which are two in number, and the whole of the head is covered with a multitude of tiny hairs of a light golden colour.

The bee has, of course, a brain in the proper sense of the word; it is, however, very minute, though all the more wonderful for being so. The nervous system consists of a number of "nerve centres," which are situated in the body. The chief nerve centre, or *ganglion* as it is called, is in the head, and from this point multitudes of nerves run to all parts of the body. The word ganglion comes from the Greek, and means a knot, and it is really a knot of nerves. The nerves resemble underground telegraph wires, which perhaps you have seen; and like them, they run in bundles, which in turn are enclosed in a pipe or sheath. Each telegraph wire sends a message to some part of the country, and the nerves of the bee, in like manner, transmit messages to different parts of its body. Other ganglia are situated in the thorax and in the abdomen, but the largest one is, as I have said, in the head. You will easily understand from this, that the ganglia are almost like little brains, distributed in the body of the bee. Now here is a most remarkable fact, but perfectly simple when you understand what I have just told you. Sometimes a bee may have a fight with another bee, and perhaps she will be unfortunate enough to have her head cut off. You might imagine that this would be at once fatal to the bee, but it is not so. She is still able to walk about the hive in quite an important fashion! Of course she cannot see, nor can she feel her way about with her antennæ, and she is therefore of no use. Soon she will die, but the fact remains that a bee can live for a time even when its head is cut off. In the same way, if a bee is feeding on honey and her abdomen is cut off at the waist, she will still go on sucking up the honey, in blissful ignorance of the fact that her body has been cut in half! Then if the abdomen is picked up and placed in the palm of the hand, it will probably start twisting round, in the attempt to bury its sting in the flesh!

CHAPTER VII
THE WONDERFUL ANTENNÆ

WONDERFUL as all the parts of the bee are, there are none so wonderful as the *antennæ*. This word comes from the Latin, and means horns or feelers, and the antennæ serve many purposes. In the hive, although all is dark, the bees are able to find their way about by means of them; they build the combs by their aid, and with them they communicate one with another. The antennæ are used, too, for the purpose of smelling, and curious to relate, the ears of the bee are situated in them. We generally expect to find the ears of living creatures in their heads, but in the insect world ears are found in many queer places. For instance, who would look for the ears of the cricket in one of its legs? yet this is where they are situated. This is not the only insect which has its ears in its legs, for those of the grasshopper are found in a similar position. Then there is a kind of shrimp, called the *Mysis*, and this creature actually has its hearing apparatus in its tail! And so, when we remember these peculiarities, the fact that the bee's ears are situated in its antennæ is not so strange as it at first seemed. In (*b*) Plate VI. you will see the position the antennæ occupy on the worker bee's head, whilst (*a*) Plate VII. will show you the feeler in detail. The antennæ of the worker bee each consist of a single long joint, and eleven small joints. The long joint is called the "scape," meaning a shaft or stem, whilst the small ones are called the *flagellum*, a Latin word meaning "a little whip." In (*a*) Plate VII. they have been numbered 1 to 11, as you will see. The antennæ of the drone, while resembling those of the worker, have one more small joint in the flagellum, thus making the total number twelve.

PLATE VII

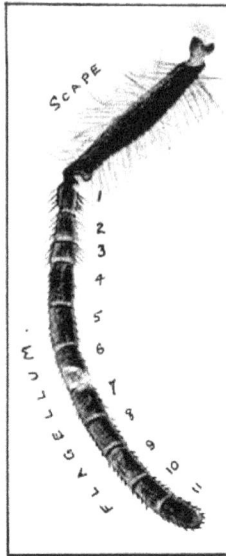

(*a*)

Photo-micro. by] [E. Hawks

Antenna of Bee

(*b*)

Photo-micrograph by] [E. Hawks

Tongue of Bee

The construction and movements of the antennæ closely resemble those of our own arms, the flagellum corresponding to the forearm, whilst the scape is like the upper part of the arm, between the elbow and the shoulder. Further than this, the antennæ are fixed to the head in much the same way as our arms are joined to our shoulders. This joint is called a cup-and-ball joint, and it enables the antennæ to be moved in practically every direction. In addition, each of the eleven joints of the flagellum is able to be moved separately; so you will see that a bee can very easily and quickly place its antennæ in almost any position.

On again looking at the plate, you will observe that the scape is covered by numerous hairs, which are both long and fine. The first three joints of the flagellum are also covered with hairs, which, however, are not like those of the scape, for they are much shorter and thicker. They look more like bristles, and all point in a downward direction. The remaining eight joints are covered with multitudes of still smaller hairs, and these again differ in their construction. To give you some idea of the complicated nature of the antennæ, I may tell you that the drone possesses over 2000 of these hairs on each one, whilst the worker has about 14,000. Each hair is connected with a nerve which is so delicate that the faintest touch of anything would be easily felt. The nerves are contained in the central part of the antennæ, which is hollow, and from there they lead to the ganglia. The bee can tell instantly the shape, height, and nature of any object by simply passing the antennæ over it. You know that if a person comes noiselessly behind you, say whilst you are reading, and lightly touches one of your hairs, you can feel the touch instantly. That is because each hair, like those of the bee, is connected with a nerve. You will easily understand, however, that the hairs and nerves of the bee are infinitely more sensitive than ours. It is necessary that the tiny workers should be provided with some means of doing things in the dark, for all the work of the hive has to be done under these conditions. The antennæ serve this purpose perfectly.

In a very powerful microscope it is found that the places between the hairs, in most of the antennæ joints at any rate, are covered with tiny oval-shaped holes and depressions. The nature and use of these holes are most difficult for us to understand, and it is not yet properly known for what they are really intended. In the first place, they are so very tiny that we can hardly imagine their size. They measure only about $\frac{1}{10,000}$th part of an inch across, and each is surrounded by a minute ring of a bright orange colour. It is supposed, and I think it is quite probable, that by the aid of these holes the bee hears. There is not the slightest doubt that bees can hear, though at one time people had quite decided that they were perfectly deaf!

In addition to these little hearing holes, there are others called the "smell hollows"; they too are exceedingly numerous and minute. Each of the last eight joints of the worker bee's antennæ is stated to have fifteen rows, and twenty smell hollows in each row! That is to say, there are over 2400 in each antenna. The queen has not quite so many, having, as a matter of fact, about 1600 on each; but the drone is possessed of the most of all, and his number reaches the astonishing figure of 37,000 hollows on each antenna. Every one of those hollows is a little nose, so that the bee's power of smell must be very keen. What with the different kinds of hairs, so numerous and yet each with a separate nerve, the hearing holes, and lastly the smell hollows, you will, I feel sure, agree that the antennæ are most complicated, and you will understand why I call this chapter "The Wonderful Antennæ."

CHAPTER VIII
THE EYES

THE same tiny head, which carries the marvelous antennæ, is provided with two large "compound" eyes, as they are called. If you are able to examine these eyes with a magnifying glass, you will at once see that they are lovely objects. The eye itself is of a deep purplish-black colour, and has an appearance which is rather difficult to describe. It seems almost as though it is covered with the finest satin, for it glistens in the sunlight.

The microscope shows that this appearance is due to the eye being composed of multitudes of six-sided cells, resembling, in fact, nothing so much as a piece of honeycomb. These cells are called *facets*, which means "little faces," and each one measures about $\frac{1}{1000}$th part of an inch in diameter. Over the surface of the eye are distributed numerous long, straight hairs; the chief purpose of these hairs is to protect the delicate facets, just as the eyelashes of our own eyes protect them. Bees have no eyelids, as we have, and so they have to rely upon these hairs to protect their eyes from dust and other such foreign bodies. The construction of the eye itself is wonderful to a degree, but it is also very difficult to understand, because it is so complicated and minute.

Each eye consists of a great number of facets, which are really smaller eyes, and this is the reason the eye is called compound. The eye of the worker contains over 6000 of them, and each one points in a slightly different direction. Large as this number may appear, it is less than half that possessed by the drone, whose facets actually number 13,000 in each eye. As a matter of interest, I may tell you that the queen bee has the least number of all, having but 5000. Each facet acts as a tiny lens. A lens, as you perhaps know, is something so shaped as to throw an image of the object to which it is directed. A camera has a lens of glass, and by the aid of this lens a picture can be taken of any object to which the camera is pointed. In that case the image of the object is thrown upon what is called a photographic plate. Our own eyes act as lenses, and throw an image of whatever we look at, not upon a photographic plate, but upon a sensitive surface called the *retina*. This word comes from the Latin, and means a "small net," and it is a very good name, for the retina catches the picture from the pupil of the eye, and passes it on to the brain.

Although we might imagine that these compound eyes were sufficient for any purpose, yet we find that the bee has three more eyes; these are called the "simple" eyes. They are situated on the top of the head, and you may see one of them in (*b*) Plate VI. The other two are over the top of the head, for the three eyes are arranged in this manner ∴ so as to form a triangle. You will remember that the drone is furnished with a far greater number of facets than the worker. Consequently the compound eyes of the drone are much larger, and they not only take up the whole of the space at the sides of the head, but also extend right over the top, covering the position occupied by the simple eyes in the worker. Owing to this fact, the drone's simple eyes are placed lower down, on the front of his head, their position corresponding pretty closely to the place our own eyes occupy. The simple eyes are so called because they do not seem to be nearly so complicated in their construction as the compound eyes, but the microscope shows that they also have an elaborate structure. If we were to cut open the front of a bee's head, we should find that the simple eyes are set like this:—

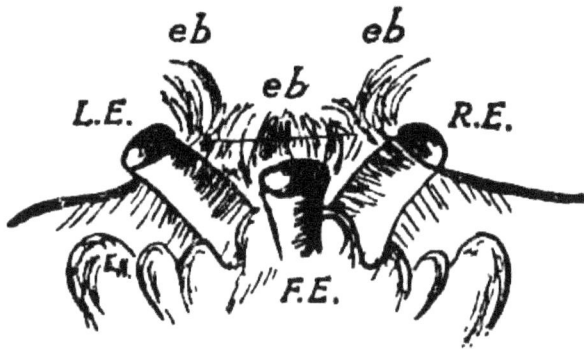

You will notice that the two top ones (marked L. E. left eye and R. E. right eye) point in an outward direction, and it is by their aid that the bee can see sideways. The lower eye (F. E. front eye) is directed forwards, and with it things in front can be seen. The simple eyes are surrounded with tufts of hair (marked e. b. eyebrows), which are so placed that they do not interfere with the range of vision.

I must just tell you something of the uses of the five eyes. At one time it was supposed that *each* facet of the compound eyes made a separate image of the object to which it was directed. But this is very improbable, for what possible use could there be in the insect seeing, instead of the one flower at which it was looking, several thousands of flowers each exactly like the other? It is much more likely that every facet forms a picture of only that part of the object which is exactly in front of it, all the pictures combining to form a single image. No doubt the compound eyes are used for seeing things at a distance, and the simple eyes for objects near at hand.

It has been proved that bees can distinguish between colours, and even that they prefer certain colours to others; one of their favourite colours is pale blue. An experiment, which is both interesting and instructive, has often been performed, and it shows us that not only is the bee able to tell one colour from another, but also that it possesses a memory. Pieces of blue, yellow, and red paper are obtained, and upon each is placed a slip of glass. A little honey is placed upon the slip of glass which is over the blue paper, and all three are put near a hive. A bee is caught and placed on the honey. After sucking some of it she flies to the hive to store her treasure and quickly returns for more. She is allowed to make several journeys between the honey and the hive, so as to impress upon her memory that the honey is to be found on the blue paper. Then while she is away at the hive, the slip of glass is placed upon the yellow paper. She returns, as before, to the blue paper, and seems puzzled at not finding the honey there, but after a careful search, she discovers the honey on the yellow paper. The fact that the bee came back to the blue paper proves that she has a memory and that she is able to distinguish one colour from another.

CHAPTER IX
THE TONGUE AND MOUTH PARTS

THE tongue of an insect is called the *proboscis*, a Greek word meaning a front feeder, or trunk, and indeed the bee's tongue is not unlike the trunk of an elephant. Let us glance at Plate VIII., where a picture of the mouth parts of the bee is shown. The tongue itself is in the centre, and it appears long and hairy, tapering to a fine point. On each side of the tongue are the *Labial palpi*, which are part of the case in which the tongue is kept, when not in use. Beyond these are the *Maxillæ*, or inner jaws, which form the other part of the case.

Each labial palpus consists of four joints, the upper two (Nos. 1 and 2 on the picture) being much larger and broader than the lower ones, which are quite tiny in comparison. They have several hairs growing upon them, and these hairs are used for feeling. The importance of hairs to the bee is very great, and we find them all over the body. They are of different shapes and sizes, and we shall read more about them as we come to consider each kind in turn. When the labial palpi are closed, they protect the back part of the tongue, the front part being protected by the maxillæ. These four parts, when closed, make a kind of tube, in which the tongue rests. Although this protecting case cannot be drawn up into the mouth, the bee is able to draw up the tongue at will.

PLATE VIII

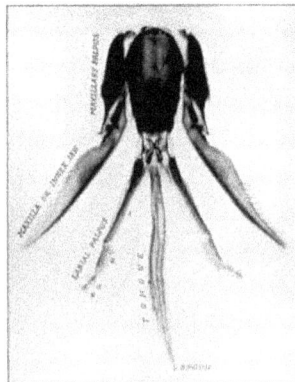

From a photo-micrograph by] [E. Hawks

Tongue and Mouth Parts of Bee

(*b*) Plate VII. shows a good view of the tongue itself, as seen with a high magnifying power. It is composed of a number of ring-like structures, and is covered with hairs which are regularly placed and point in a downward direction. The tongue of the worker bee, it is interesting to note, is nearly twice as long as that of the queen or of the drone. This is because neither of the latter gather nectar, and so they do not need such long tongues as the worker. Her tongue being longer, she is the more easily able to reach the nectar, which, in some flowers, is only to be found at the bottom of a long corolla. The tongue of the worker has from 90 to 100 rows of hairs, but those of the queen and the drone have only from 60 to 65 rows each.

The tongue is extremely elastic, and is capable of being moved in any direction at will. Some of the hairs with which the tongue is clothed are of use for feeling, but most of them are for a different purpose altogether. When a bee pushes her head into the corolla of a flower, her tongue sweeps from side to side. If there is any nectar there, it sticks to the hairs of the tongue in tiny droplets, and in this way it is collected. Later on we shall find how it is dealt with after it has been gathered.

On (*b*) Plate VII., at the very tip of the tongue, there is to be seen a small object like a spoon. This is indeed its name, and it is used for collecting the most minute quantities of nectar. It is covered with a number of tiny hairs, some of which are split into several branches.

From this description you will see that a bee's tongue is very fully equipped for gathering small, as well as large, quantities of nectar. Even the tiniest drop is carefully treasured, for the bees know that "every little helps."

CHAPTER X
THE JAWS

WE have seen that the bee possesses maxillæ, or inner jaws, and we are now to consider the outer jaws. On (*a*) Plate IX. is a photograph showing these jaws, which have been separated from the mouth in order to show them better. They are very hard, and have extremely sharp edges, like a joiner's chisel. If you have ever watched a caterpillar feeding, you will know that its jaws work sideways. It places itself upon the edge of a leaf, and moves its jaws from left to right, one on each side of the leaf. This action therefore resembles the opening and shutting of a pair of scissors, placed flat upon the table, and the jaws of all insects work in a similar manner.

The jaws of the bee are very powerful, and this is necessary, for it is by their aid that the wax, which forms the comb, is cut up or thinned out. Sometimes the bee may come to a flower which is too long for its tongue to reach the bottom. It does not waste time trying, but simply bites through the flower, inserts its tongue through the hole, and in this way obtains the nectar.

A short time ago I imprisoned a wild bee in a cardboard box. Soon afterwards I heard a great noise coming from the inside of the box, and found that the little captive was hard at work, endeavouring to bite a way through the cardboard. The noise made by its tiny jaws, as it tore away shred after shred of cardboard, was like a mouse gnawing a plank. I fed the bee with honey, and the next day found the floor of the box covered with pieces of cardboard, whilst quite an appreciable amount had been bitten away. In four days the bee had cut a way through the side, making a hole large enough for herself to pass through. Seeing that she had worked so hard, for the box was really a very substantial one, I rewarded the little worker by setting her free.

PLATE IX

(*a*)

Photo-micrograph by] [E. Hawks

The Jaws

(*b*)

Photo-micrograph by] [E. Hawks

Claws, showing Hooks and Feeling Hairs

CHAPTER XI
THE THORAX

HAVING now fully considered the head of the bee, we will turn our attention to the *thorax*; this name comes from a Latin word meaning the chest. It is the second, or middle division, of the bee's body, and to it the head is joined by a thin neck. The *thorax* is the centre of movement, for it is to this part that the wing and legs are joined. Accordingly we find that it contains several large muscles, for the bee is a very powerful flier.

If we examine a bee we notice that the head seems almost black, the abdomen smooth and shiny, and that the thorax has a beautiful downy appearance. This is due to its being thickly covered with fine hairs which, when examined with the microscope, are seen to have many tiny spikes branching from them which are used for collecting the pollen grains. When a bee enters a flower the hairs are sure to come into contact with the pollen, and by means of the spikes the grains are entangled and held secure. The hairs of the queen and the drone are not so numerous as those of the worker, because these bees do not gather pollen.

If we wish to see exactly the construction of the thorax we shall have to remove these downy hairs, for they are so thick that it is impossible to see beneath them. How are we to remove them, without injuring the parts which lie below? An ingenious way of doing this is to fasten a piece of cotton around the body of the dead bee, and to hang it downwards in the hive, between the combs. In the course of a few days we shall find that every hair has vanished and that the body is beautifully polished. This has been done by the thousands of worker bees, walking over the combs of the hive. They are so busy that they have no time to stop and inquire how their sister died; and so they brush past, intent only on the fulfilment of some particular duty. In their haste they knock against the body of the bee, which is buffeted this way and that, as the busy streams of bees cross and recross the combs. After a few days of this treatment all the hairs will have been removed from it, and we shall then be able to see the actual construction of the thorax, and also the manner in which wings and legs are attached.

The thorax, we find, is divided into three distinct parts. The division nearest the head is called the pro-thorax or forward division; the second is the meso-thorax or middle division; and the third the meta-thorax or after division.

CHAPTER XII
THE LEGS

THE legs of the bee are not only used for walking but they have also to take the place of hands and arms. They are divided into three pairs, one attached to each division of the thorax. Each leg has nine joints, which have separate names. The last joint, which is really the foot, has two claws and a kind of soft pad. The claws, a picture of which is shown in (*b*) Plate IX., are useful for walking over rough surfaces, and also serve as little hooks. When the bees are wax-making they hook their feet together, just as we take hold of hands, and they are thus able to hang in long festoons from the roof of the hive.

The pad is called the "pulvillus," and is close to the claws. We all know how easily a fly can walk upside down on the ceiling, or run up a window pane. It is able to do this by means of pads which it also possesses. These pads are covered with a kind of gummy liquid, and by their aid a fly or a bee can walk up, or perhaps it would be more correct to say stick to, a window pane or other smooth surface. The fly, however, can beat the bee when walking on such surfaces, because it has two pads on each foot, whereas the bee has only one. On the other hand, the claws of the fly have no hooks, therefore flies cannot cling to each other as bees do.

It is very interesting to understand how the pads are brought into use by the bee. You must remember that they are placed just above the claw itself; when the bee is walking over an uneven surface the claw catches on the roughnesses, and then the pad remains in its ordinary position. When the bee comes to a slippery surface, however, the claw is not able to obtain a grip, and so it slips down under the foot, its place being taken by the pad. This presses against the smooth surface and adheres to it by means of the sticky moisture with which it is covered. Here is a sketch showing the pad just coming into action.

- 25 -

The pads hold very tightly on to a smooth surface when they are pulled *downwards*, as it were, by the weight of the bee. But they are very easily loosened if the sides are lifted up, and in this manner they may be peeled off the smooth surface, just as we take a stamp off a letter. So beautiful is this arrangement, and so perfect in its action, that it is stated a bee can put down and lift up each foot at least 1200 times a minute!

CHAPTER XIII
THE FIRST PAIR OF LEGS

THE first pair of legs, or those nearest the head, are the shortest of all. The most interesting feature about these legs is a little semi-circular notch, and I have made this sketch of it.

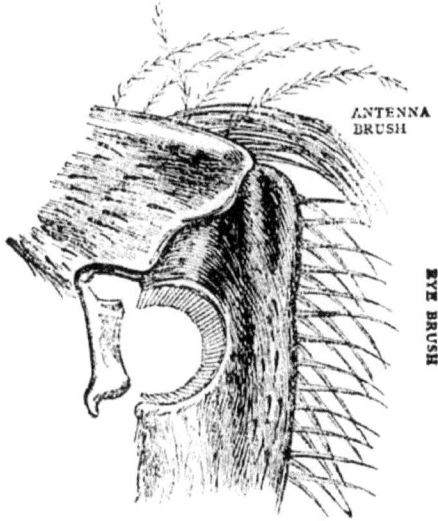

Under the microscope we see that around the semi-circular opening is a row of about eighty teeth. These are not biting teeth, but are more like the teeth of a comb, and indeed this notch is a comb which is used for cleaning the antennæ. You may sometimes see a bee bring up its front leg to its head, and then move the leg outwards. By this movement the antenna is drawn into, and through the comb, the teeth of which soon remove any dirt or pollen which may be sticking to it. Just above the antenna comb, there is a kind of little hinge or lid. This is called the "velum," and its name comes from a Latin word meaning "to cover," for the lid covers the antenna when it is drawn into the comb, and holds it there whilst it is being pulled through. When we know that each antenna is only $\frac{1}{125}$th of an inch in diameter, we can understand what a wonderful little tool the comb is.

When we mention a comb, we generally think of a brush too, so it is interesting to find that the front leg of a bee has two brushes, which are shown in the sketch. The first of these is used for cleaning the comb after the antenna has been passed through it. The other keeps the hairs of the eye free from pollen.

PLATE X

(*a*)

From a photo-micrograph by] [E. Hawks

Hind Leg of Bee (showing Wax Pincers)

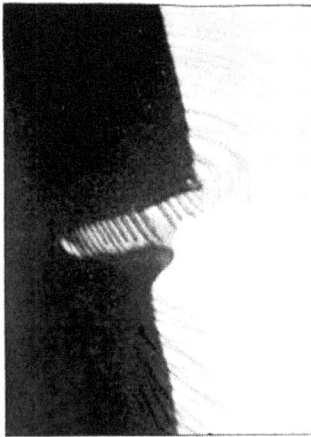

(*b*)

From a photo-micrograph by] [E. Hawks

Wax Pincers on Hind Leg

CHAPTER XIV
THE SECOND AND THIRD PAIR OF LEGS

THE second pair of legs is slightly longer than the front ones. Each is furnished with a kind of stiff spike with which the wings are cleaned.

The third pair of legs are perhaps the most interesting of all. They are the longest, and the hairs for pollen gathering are far more numerous upon them than on the other legs. If we look at (*a*) Plate X. we see that there is an opening in the leg around which is set a row of spikes. This is shown more plainly in (*b*) Plate X. As the joints work on a kind of hinge, these spikes act like pincers; they are known as the wax-pincers and will be mentioned later. Another interesting feature is the *corbicula*, or pollen basket, which is the receptacle in which the pollen is carried from the flowers to the hive. You will see from the picture of the worker bee, in Plate V., that the large joints of the hind legs are much broader than the others. They are also hollowed out, and around each edge are numbers of spike-like hairs, which curl inwards over the hollow. These make a sort of basket, and I am sorry that I am not able to show you a photograph of this interesting feature, but it is a most difficult subject of which to obtain a picture. However, I have made this little drawing, which perhaps will help to give you some idea of its nature.

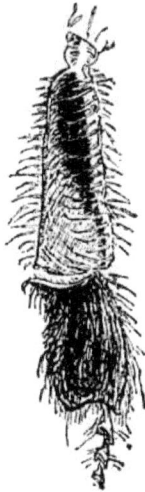

POLLEN BASKET

POLLEN BASKET

I should tell you that the pollen basket is situated on the outside of the leg, that is, the side which is away from the bee's body. On the inside are several combs, which are made up of rows of spike-like hairs. When the thorax has become covered with pollen the bee uses these hairs to comb it out; this it does by crossing its legs below the body. It is interesting to notice that neither the queen nor the drone has pollen baskets.

PLATE XI

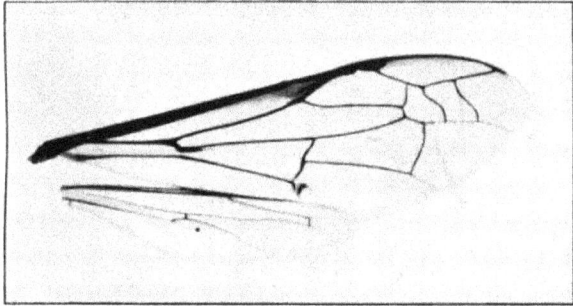

(*a*)
From a photograph by] [E. Hawks

Wing

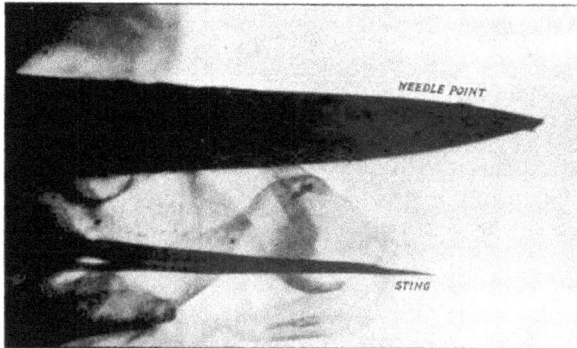

(*b*)
From a photo-micrograph by] [E. Hawks

Fine Needle compared with Sting

CHAPTER XV
THE WINGS

BEES belong to a class of insects known as *Hymenoptera*, which means with membranous wings; the wings of the bee are found to be composed of beautifully fine membranes. They are four in number, and, like the legs, are joined to the thorax. The front ones are called the anterior wings, and the back ones, which you will notice are smaller, are called the posterior wings, because they are behind the others. The membranes are strengthened by a kind of framework, just as a kite is strengthened by a framework of light sticks. The ribs of the framework are called "nervures," and, as you will see from (*a*) Plate XI., there are divisions of transparent membrane in between; these are called cells. The nervures are hollow, and like our veins, they contain blood.

We have seen that the bee possesses two pairs of wings, and we may wonder why this should be so, when we know that one large pair is much more powerful for flying purposes than two small pairs. You have no doubt noticed that when a bee is at rest on a flower the wings are neatly folded over the back. Now if the bee had only one pair of large wings it would not be able to fold them so compactly—the wings would, in fact, stand out on each side of the body. We shall presently see that the bees, in the course of their duties, have to clean out the cells of the comb, and in order that they may do this it is necessary for them to be able to crawl right into the cell itself. The cells in which the young worker bees are raised are only $\frac{1}{5}$th inch in diameter, and if the wings projected when in the folded position, the bee would not be able to enter the cell. The wings therefore have been divided, so that when folded they may lie one over the other on the bee's back, and we find that the wings, when folded, take up only $\frac{1}{6}$th inch of room. This leaves just sufficient space for their owner to enter a cell. You will notice that a blue-bottle fly has only one pair of large wings, for it does not need to fold them closely over its back, as it has no cells to clean.

Remembering what I have told you about the greater flying power of one pair of large wings, you might imagine that the division into two pairs which we have seen to be necessary would handicap the bee in flying. The difficulty is overcome by a most ingenious device, by which the bee, when flying, is able to fasten together the wings on each side, so as to form one pair of broad wings.

Let us now turn to (*a*) Plate XII., which shows part of the wings on one side of a bee's body. Along the top edge of the lower wing there is a row of tiny hooks, and the lower edge of the upper wing is curled over, thus forming a kind of ridge. When the bee takes to flight the front wing is stretched out from over the back, and during this action it passes over the upper surface of the back wing. When the ridge reaches the hooks it catches upon them and is held fast. In this manner the two wings are locked together. (*b*) Plate XII. shows the wings hooked together ready for flying. When the bee comes to rest she folds her wings, and in doing this they are automatically separated, for the ridge slips away from the hooks that hold it.

PLATE XII

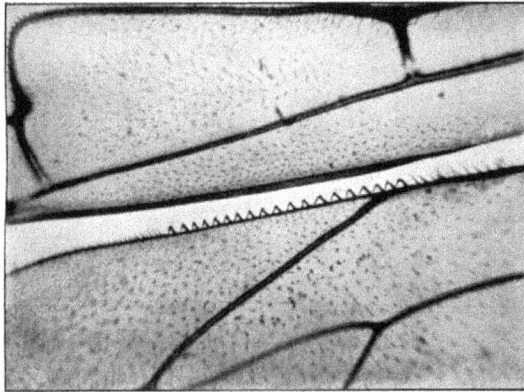

(*a*)

From a photo-micrograph by] [E. Hawks

Wing unhooked, showing Hooklets and Ridge

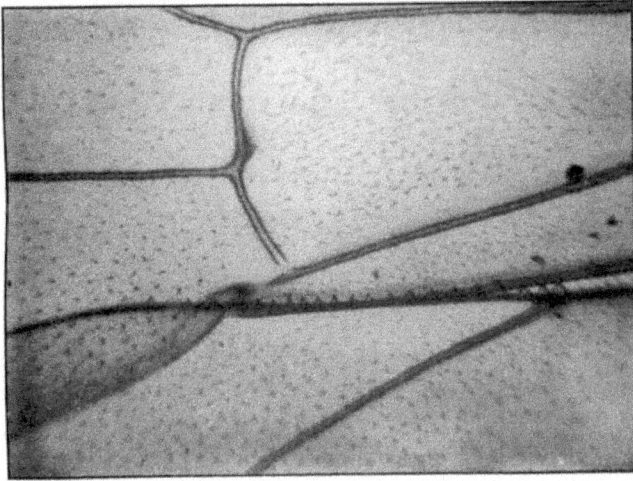

Wing hooked, as in Flying

The number of hooks varies, and there are sometimes more on one side of the body than on the other. As a general rule it is found that a worker bee has from eighteen to twenty-three of them, the one shown in (*a*) Plate XII. having nineteen, as you will be able to count. The queen does very little flying, and so her wings are not large, in proportion to her size. Therefore she has not usually so many hooks, and sometimes they are found to number as few as thirteen. The drone has large and powerful wings, and his hooks vary between twenty-one and twenty-six in number.

Bees are able to move their wings very quickly, and you will agree with me in this when I tell you that it has been shown that the vibrations number at least 190 per second! The flight of the bee is greatly assisted by a number of air-sacs called *tracheæ*, contained in the thorax. These fill with air and make the body more buoyant, just as a lifeboat is made more buoyant by its air-chambers. When a bee has been at rest for a little time it cannot begin to fly straight away, for the air-sacs are empty. It therefore runs along the ground to get a start, as an aeroplane does, and by vibrating its wings fills the tracheæ.

CHAPTER XVI
THE ABDOMEN

THE hinder part of the bee's body is called the abdomen, and it is here that the stomach is situated. The abdomen is larger than either the head or the thorax, and is joined to this latter by a thin waist. Insects do not possess skeletons, at least not internal skeletons of bones, such as we have. Their skeletons are outside the body, and take the form of a hard outer layer which protects the soft inner organs. This layer, or outer skin, is made of a horny substance, called *chitine* (pronounced "ki-tin"), which comes from a Greek word meaning a tunic or outer dress.

Chitine is indeed a wonderful substance, and is found in all forms and shapes, having a variety of appearances. The hard black bodies of beetles are composed of it, and, wonderful to relate, of this substance the downy wings of the butterfly are made. You will remember that in the chapter on the eye of the bee we saw that the facets have a beautiful appearance; they too are made of chitine, as are the tendons, legs, hairs, membranes, and many other parts of the body.

The abdomen of the queen and of the worker is divided into six rings or belts, but the drone, having a somewhat larger body, has seven. Each ring is divided again into two parts which are known as the *scelerites*, which are joined one to another by delicate membranes of very fine skin. You may have noticed that the leg of a crab is jointed, and that the hard outer case of shell gives place to a fine, but tough membrane at the joints. By means of this arrangement the crab can move its leg with ease. The joints of the abdomen of the bee are arranged in a similar manner, although in this case the membranes are of course much finer and more delicate than those of the crab.

The organs inside the outer case of chitine are of most wonderful and delicate construction. You may be surprised and interested to learn that a bee has two stomachs, and these are perhaps the most important parts of the abdomen. It is not because the bee is a greedy insect that it is provided with two stomachs, but each serves a separate and useful purpose. One is called the honey-sac, and the other is the stomach proper. As a bee sips the nectar from a flower, it is passed down a tube through the thorax into the honey-sac, which acts as a kind of store-chamber. Here it is kept until the bee flies back to the hive, or until the little worker may need it for its own food. Leading from the honey-sac to the stomach is a very fine tube, and at the honey-sac end of it there is a kind of stopper, called the "stomach mouth." Just as we can open or close our mouths at will, so can the bee open or close the stomach mouth, and so either allow honey to flow into its true stomach or keep it stored in the honey-sac. The latter is very tiny, and when quite full contains little more than a third of an ordinary drop of honey. The tube which leads from the one to the other is lined with fine hairs, all pointing in a downward direction, away from the honey-sac. When the bee sips the nectar it often happens that some of the pollen grains from the flower are taken in also. Now the bee desires to gather only the pure nectar, and so it passes the nectar from the honey-sac to the stomach by means of the tube. It then makes the honey return from the stomach to the honey-sac, but this time the hairs in the tube act as a strainer, and prevent the pollen grains from returning with the nectar. By this clever little apparatus you will see that the bee is able to strain the nectar when flying from one flower to another, or when travelling back to the hive. Besides the two stomachs, the abdomen contains certain glands to which we shall refer when we come to speak of honey.

CHAPTER XVII
THE BREATHING APPARATUS

INSECTS do not breathe by means of lungs as we do but through tiny air-holes, called "spiracles." This name comes from the Latin *spiraculum*, meaning an air-hole, which in turn is derived from *spirare*, to breathe.

Crawling insects do not need nearly so much air as flying insects, and so their breathing apparatus is not so large. In the bee the breathing tubes spread over almost the whole body, two of the largest extending along each side of the abdomen. The rings of the abdomen slightly overlap one another, and if you watch a bee carefully you will notice that they are constantly slipping in and out, like the joints of a folding telescope which is being opened and closed. This is really the action of breathing, and the bee draws in and then drives out air. If you have ever rescued a fly which has fallen into the milk, you will remember that it at once commences to clean itself vigorously with its legs. It does not do this to make itself tidy, but to clean out the milk which clogs its air-tubes and is thus choking it.

It is interesting to notice that the mouth of each air-tube has a number of tiny hairs; these serve to keep out dust, which would interfere with the breathing. The air-tubes branch off one from another like the roots of a tree, and in order to give you some idea of how very small they are, I may tell you that it has been found that a bundle of a quarter of a million of them would hardly be any bigger than an ordinary human hair!

PLATE XIII

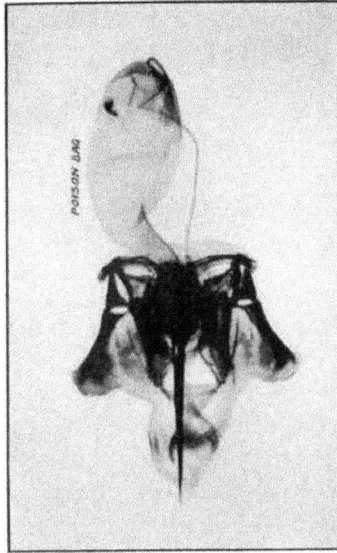

(*a*)

Photo-micrograph by] [E. Hawks

Sting of Bee

(*b*)

Photo-micro. by] [E. H.

Sting, showing Barbs

CHAPTER XVIII
THE STING

WE have now only the sting left to consider. I need not tell you what it feels like to be stung, as no doubt a good many of you have had that interesting operation performed upon you by some bee or wasp which you have annoyed!

How very frightened every one is of the sting of a bee, and those people who have never been stung are perhaps the most frightened of all. After all, the sting is not so painful, and it is very interesting to watch the angry little worker drive its sharp weapon into our hand; besides which it is actually good for us to be stung, and the reason of this I shall presently tell you. The sting is situated at the very tip of the abdomen. It would take up too much space to fully describe all the details of its construction, and therefore I shall simply tell you about the chief parts, and also how it works.

Let us look at the picture of a sting given on (*a*) Plate XIII., where is seen a sharp-pointed object surrounded by fleshy matter. This is the sting proper, and it is very smooth and hard, as well as being finely pointed. In order to give you some idea of this, I have mounted alongside a sting, one of the finest needles obtainable for comparison, and you will see the picture in (*b*) Plate XI. The needle is at the top, and looks like a great crowbar compared with the beautifully fine and tapering sting.

This sting is really a sheath, or kind of case, in which are enclosed two needle-like darts. Its purpose is to protect the darts and also to make the actual wound. Outside the end of the sheath are two rows of three, or sometimes more barbs, which point backwards. Many of you, no doubt, have seen in our museums the spears and arrows used by savages, which have ugly barbs at their points. When the warrior runs the spear into an enemy, it does not slip out as it would do were the shaft just a plain one. The barbs on the outside of the sheath are used for this purpose, that is, to prevent the sheath from slipping out of the hole it has pierced, until the operation of stinging is completed.

The darts enclosed in the sheath are capable of being moved up and down in it, by a powerful and complicated set of muscles. They act like drills, and when the sheath has made the first hole and, as it were, opened the way for them, the darts commence to travel up and down at a great rate. Every time they come down they go further into the flesh, and so make the hole deeper. They, too, have barbs which are more pronounced than those on the outside of the sheath, and so take a firmer hold on the flesh. You will clearly see these barbs on one of the darts in (*b*) Plate XIII.

The darts themselves are hollow, and near each barb there is a tiny hole, which leads into the central hollow, down which the poison is poured. The hole made by the sharp little darts is not deep enough to cause the pain we feel when stung; this is due to the poison which is sent into the wound. This poison consists chiefly of formic acid, and is stored in the poison-bag which is shown on (*a*) Plate XIII. The poison is forced through the holes by two little pumps situated at the base of the sheath, and which are worked by the same muscles which move the darts.

You will see from this that stinging is quite an elaborate process. First the sharp point of the sheath enters the flesh and is held there by its barbs. Then the darts work up and down, making the wound deeper and deeper, while the tiny pumps are forcing in the poison. So quickly does all this take place that the sheath is driven in up to the hilt and the wound filled with poison, long before we have time to knock the angry little insect away.

When a bee stings our arm or leg we naturally try to brush or shake it off. We have seen that the sheath of the sting has barbs, and when we shake our arm the sting is so fast in the flesh that the jerk causes it to be pulled out by the roots from the bee's body. When this occurs it generally happens that a large part of the bee's bowel is pulled out also, and this causes the death of the bee in an hour or so. If we let the bee alone, however, we shall find that after the darts have been driven in as far as ever they will go, and after the full amount of poison has been pumped in, she will commence to turn slowly round and round, and in this manner will extract the sting, as a corkscrew is taken out of a cork.

The sting of a worker is quite straight, but that of the queen is curved like a scimitar. The workers sometimes sting bees from other hives, but the queen will never sting any bee but a rival queen. The sting of one bee is immediately fatal to another.

CHAPTER XIX
THE ANCIENTS AND BEES

BEFORE we go on to consider the habits of the bees, I think you will be interested to hear something about their early history, and how they used to be kept in bygone ages. Thus we shall be able to trace the progress of bee-keeping from its earliest sources to the present day, and to realise the wonderful improvements of modern methods upon those of the ancients.

It is not possible for us to tell with any certainty when bee-keeping actually commenced, but it has a very ancient origin. No doubt for ages past it has been the custom of men to obtain honey from the store of wild bees. For instance, we read in the Bible that John the Baptist lived for some time in the wilderness on locusts and wild honey. The earliest records in existence show us that the Egyptians kept bees in some kind of hive, and that they carefully studied their habits. If you visit the Egyptian rooms at the British Museum, you may perhaps see the sarcophagus which contains the mummified remains of a great king, called Mykernos. This coffin dates back to 3633 years B.C., and Mykernos was at that time the King of Lower Egypt. On the outside of the coffin is a peculiar drawing, or hieroglyphic as it is called. It is something like this:—

This funny little figure represents a bee, for at that time it was thought that the bees were ruled over by a king-bee, which the Egyptians knew to be larger than all the others. Because the bees always appeared to be so happy under their king, the Egyptians thought it would be a good symbol to place on the coffin of their ruler. This is the very earliest known record relating to bees, but we know now, of course, that the large bee, which seemed to the Egyptians to rule the others, is not a king but a queen.

Those of you who learn Latin may some day have to translate some books called the *Georgics*. They were written by a clever man called Virgil, and although schoolboys do not always like them, yet they are most interesting, especially the Fourth Book, which tells us a great deal about bees. Virgil lived in a town called Parthenope, which we now know as Naples. He was a great bee-keeper, and was never tired of watching his bees at their work, and moreover he left very accurate accounts of his observations. Hives in those days were dome-shaped, and made from pieces of bark stitched together, or sometimes of osiers or plaited willows. We can imagine the learned Virgil walking in his garden, surrounded by sweet-smelling flowers and herbs, and by his quaint bee-hives. Below, down the mountain side, lay "sweet Parthenope," as he called it, with its orange and lemon groves. Beyond the town lay the most beautiful bay in the world, the Bay of Naples, whose water, as blue as turquoise, shimmered in the summer sun. Over all stood the crater of mighty Vesuvius, from the cone of which a thin wisp of smoke hung lazily in the atmosphere. In this way Virgil spent many happy days, and in the book I have mentioned we may read of his doings, and of his bees. Most of his ideas about bees were false, but some of the rules which he laid down for bee-keeping hold good even at the present time.

Up to the time of Virgil, and even later, the duties of the workers in the hive were not properly understood. It was not known even that the largest bee was really the mother of them all, and that the workers looked after and tended the eggs, which later on would develop into young bees. In the days of Virgil it was supposed that bees were born in flowers, or that if an ox was killed and left to decay, a swarm of bees would be formed in its body and could then be put into a hive. In the Fourth Georgic very careful instructions are given by Virgil as to how to prepare an ox for this purpose. Many years ago this was translated into our language by a bee-keeper, and the wording is so quaint that I think you will be interested to read the following extract from the curious directions. We are told that we must find "a two-year-old bull calf, whose crooked horns be just beginning to bud. The beaste, his nose-holes and breathing are stopped, in spite of his much kicking! After he hath been thumped to death, he is left in the place, and under his sides are put bits of boughs and thyme and fresh-plucked rosemarie. In time the warm humor beginneth to ferment inside the soft bones of the carcase, and wonderful to tell there appear creatures, footless at first, but which soon getting unto themselves wings, mingle together and buzz about, joying more and more in their airy life. At last they burst forth, thick as raindroppes from a summer cloude...."

The supposition that bees were obtained from a dead ox lasted right down to the seventeenth century, and there is no doubt that the Egyptians believed in this too, for in some of their records we find that they buried the body of an ox, leaving the horn-tips just above the soil. After it had been left so for about a week, the tips of the horns were sawn off, and a swarm of bees issued, like smoke from a chimney. What a foolish idea this was, just as though the body of an ox could, in any manner imaginable, change into a swarm of bees! It probably originated in the fact that the decaying body of an ox or other animal quickly becomes surrounded by swarms of flies, wasps, and other insects.

Up to the fifteenth and sixteenth centuries, the people had no other substance than honey with which to sweeten their food, for the mode of extracting the sweet juice contained in the sugar-cane was not known till later. Sugar-cane was actually discovered somewhere about the first century A.D. and a learned writer, Strabo by name, has told how the chief admiral of the fleet of Alexander the Great found what he called "a wonderful honey-bearing reed," whilst on a voyage of discovery to India. It was not until the fifteenth century, however, that the Spaniards set up a sugar plantation in Madeira, and extracted the juice from the cane: even then it was only the rich people who could afford the new luxury, and others had still to use honey. From these remarks, then, we can easily understand how necessary bees were to the people, and how much depended on a good honey year.

Besides using honey for sweetening purposes, the Anglo-Saxons made from it a drink called Mead. You have no doubt read of this in your history books, but perhaps you did not know that it was made principally from honey. Sometimes the juice of mulberries was added to it, to give the drink a flavour, and it was then called Morat. People who could afford to do so flavoured it with spices, or sometimes even added wine, and in this form it was used in the royal palace. In some country places old-fashioned people still make and drink mead, but it is very rarely heard of nowadays.

Bees also provided the ancients with wax, from which a sort of candle was made, for in those times there was no electricity or even gas, and so the people were very glad to be able to use the wax for lighting purposes. Nowadays, beeswax, mixed with a little turpentine, is used for polishing furniture and oilcloth.

PLATE XIV

The New and the Old

CHAPTER XX
THE HIVE

A HIVE may with all truth be called a bee-city, for in it there live thousands upon thousands of little workers. In this chapter I hope to tell you about the actual construction of this wonderful city, so that you may understand more easily the chapters that will follow.

Hives used to be made of straw, and were called "skeps." Some of these skeps may still be seen in country places, but they are rapidly being superseded by the more convenient wooden hive. The two kinds are shown in Plate XIV. The wooden hive is a kind of box made in a special way, and it is usually painted white, for this not only looks clean but also keeps out the heat of the summer sun. You will notice that, like one of our own houses, it is divided into three storeys. Close to the floor of the hive, at the bottom of the lowest storey, is the door, and this is made by cutting a slit in the wooden wall. Two little slips of wood slide in front of it, so that it can be made narrower, or even completely closed at the wish of the bee-keeper. If the bees themselves wish to close up the entrance for any reason, they are able to do so by blocking it up with wax. The top chamber of all is the roof, which is empty, and serves to protect the hive from the rain. It must, of course, be lifted off by the bee-keeper each time he wishes to look into the hive. The second chamber is a sort of extra storehouse, and it is used by the bees to store honey when the third chamber is full. This third chamber is the most important of all, for it is here that the bees live. It consists of rows upon rows of combs, some of which are storeplaces for honey, but the greater part form the nurseries where the young bees are brought up.

All the cells are built of wax, no matter whether they be honey cells or cradles, and they are constructed in wooden frames which the bee-keeper places in the hive for the purpose. In Plate XV. we see the roof and the second chamber removed, exposing the inside of the bottom chamber. The bee-man in the picture is lifting out one of these frames of combs in order to examine it. The frames are simply four pieces of wood, and are used so that the bees may not fasten their combs to the walls of the hive, for if this were done it would not be possible for us to remove them from the hive. The number of frames a hive contains depends on the size and prosperity of the bee-city, and also on the particular time of the year. If the city is a large one, and the inhabitants numerous, there may be twelve or fourteen frames, each containing thousands of separate cells. On the other hand, if the bees are few, or suffering from any disease, the frames may be reduced to half this number. Of course, the more numerous the frames, the greater is the amount of work to be done, and the more workers will be required to attend to the young bees, and to the duties of the hive. When all the frames are in position, they look something like the picture in Plate XVI.

PLATE XV

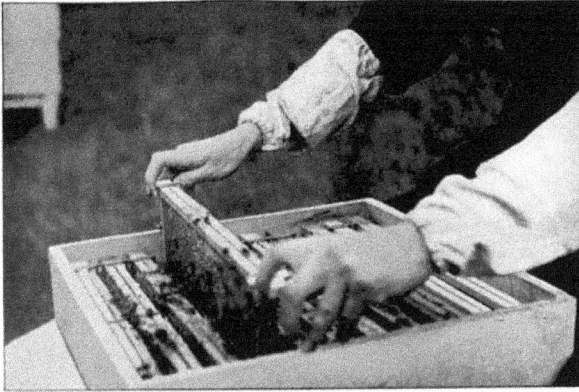

Lifting out a Frame of Comb

PLATE XVI

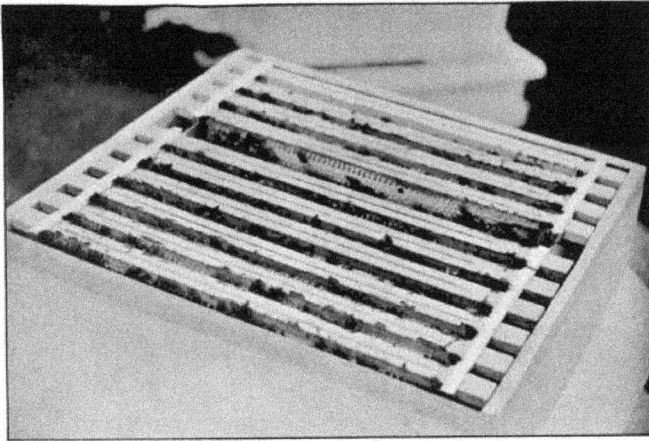

From a photograph by] [E. Hawks

Showing the Frames in Position

When we are examining a frame, we generally cover the others over with a cloth, for the bees do not like the light to penetrate their city. The frame having been replaced and the second chamber put on, we cover all over with thick pieces of felt to keep the hive warm, and on top is placed the roof. The hive stands on four legs, a few inches above the level of the ground, and the door is generally sheltered by a kind of porch. In front of the door there is a board which projects a few inches, and this is called the alighting-board. On it the bees settle when returning from the fields, and from it they commence their flight when leaving the hive.

CHAPTER XXI
A VISIT TO A HIVE

LET us now imagine that we are to pay a visit to a hive. If we are afraid of stings we may put on thick leather gloves and tie our sleeves around the wrists, to prevent any curious bee from investigating our arms. Then over our hats we may place a veil, to keep the bees from our face, for a sting in the eye would be a serious matter. The bee-man in Plate XVII. is wearing a veil, as you will see, and the brim of his straw hat is useful to keep it at a little distance from his face, so that the bees are not able to sting through it. Before we approach the hive I must tell you one thing; if a bee flies around you and comes rather closer to your face than you care about, do not on any account hit it away. Bees, like some human beings, are very curious by nature, and they like to investigate anything strange that comes under their notice. Never mind if one of them comes crawling over your hand, or even if it steps inside your ear! It will not hurt you if you keep still, but should you knock it away with your hand, it will become angry, and probably you will be stung there and then.

PLATE XVII

From a photograph by] [E. Hawks

Examining a Comb

Bees are very brave little creatures, and are frightened of nothing in the world except smoke and the smell of carbolic acid. When we wish to open the hive and to examine the combs, we must first puff in a little smoke at the door. Ordinary tobacco smoke would do quite well, but we more often use a rolled-up piece of brown paper, or some old rag, which are allowed to smoulder. They are placed inside a tin, which is fixed to a pair of bellows, and by working the bellows with our hand we are able to puff out any quantity of smoke from the nozzle with which the tin is fitted. This is done to frighten the bees, and not to stupefy them, as most people think. As soon as the smoke reaches them they rush to the storehouses in order to take in provisions, for they think some terrible calamity is about to occur. They know that they would starve if they were forced to leave the hive without a supply of food, and so by filling their honey-sacs they provide themselves with food to last at least a day or two. Though the bees are greatly frightened by the smoke, they have no intention of deserting the city that they have built with so much labour, unless it is absolutely necessary; so after taking in supplies they wait to see what is going to happen. While all this is going on we may look into the hive and examine the combs, and after doing so the roof is replaced, the smell of smoke leaves the hive, and the bees settle down again. The honey in their honey-sacs is put back into the storehouses, and work goes on as usual throughout the bee-city.

CHAPTER XXII
THE CITY GATE

THE door of the hive, or the city gate as it may be called, always presents a busy spectacle, and Plate XVIII. is a photograph of one. Bees are constantly alighting on the board, coming so quickly that they appear to spring from nowhere. Other bees come out of the gates, and fly away quite as rapidly. Some even are in such a hurry that they do not wait to crawl on to the board, before taking to flight, but fly straight out of the door and away into the blue. Then, again, others do not seem to be in such a hurry, for they come out of the gates, and stand on the board brushing down their wings, seeming almost as though they were blinking in the bright light of the morning sun. These are the young bees, who are on their first expedition to gather honey; probably they have never been outside the dark hive before, and so they are unaccustomed to the strong light. They must take careful survey of the position and surroundings of the hive, so that they will be able to find it again when returning laden with honey. The bees which dart straight off from the hive door are the older workers, who have made many a journey to and fro, and so know very accurately the position of the hive.

PLATE XVIII

From a photograph by] [E. Hawks

The City Gate

All these are the foragers, or honey gatherers, and it is their business to visit hundreds of flowers over the country side, and to extract from them, by the aid of their wonderful tongues, the tiny drops of nectar. When their honey-sac is full, they return to the hive with all speed, and rushing inside, hand over the fruits of their labours to the house bees. You will be surprised to hear that a bee has to visit over 100 flowers before her honey-sac is filled, and we must not forget that this tiny sac when full holds only one-third of a drop. Now you will understand what a great number of bees are required, and how hard they have to work, in order to make 1 lb. of honey. Yet some hives give more than 200 lbs. of honey in a season! Just think of the vast amount of labour and the incessant toil required for this result. But the bees are always busy, and the proverb, "Go to the ant, thou sluggard," might be quite well changed to "bee," for I question whether the ant really works harder than the bee. From the time that the first ray of the morning sun strikes the dewy fields, until the sunset merges into misty twilight, all is bustle and hurry in the bee-city. So hard do the foragers work that instead of living three or four years like the queen, they often live only two or three weeks in the summer. In this short time their wings become quite worn away, and their poor little bodies are covered with wounds.

If we look carefully at the door of a hive on a warm summer's day, we shall no doubt see some of these poor worn-out creatures. They can no longer take part in the great work of the hive, and so for a short time they come out into the sunshine and dodder about the alighting-board. Their mission in life being over, no doubt they will summon up all their remaining strength to fly away to some quiet spot where they will die, unheeded and unknown. Their last thought is to die somewhere away from the hive, so that their bodies may not interfere with the work of the city, and will not need others to carry them to a burial-place. How sad it is to think of these noble little workers, thousands upon thousands of which out of each hive willingly give up their lives for the great work of their race.

Besides the ever-busy foragers, there are other bees coming and going who do not appear to be in such a hurry. Each has two bright-coloured spots on her hind legs. These bees are the pollen gatherers, who collect the "bee-flour"; we might rightly call them the millers of the hive, and a picture of them is shown in Plate XIX.

Some of the bees at the city gates are employed in quite a different manner; they do not fly afar in search of honey or pollen, but stand still, with heads pointing to the hive door. They are using their wings so vigorously that we cannot see them, just as the propeller of an aeroplane is invisible, because it is turning so quickly. These are the ventilating bees, whose duty it is to keep the hive cool on hot days. The quick fanning of their wings draws out the heated air from the hive, and if we were able to peep inside the door we should see other bees also engaged in the same occupation. These, too, stand with their heads towards the hive door, but instead of fanning out the hot air, as the outside bees do, they draw a stream of pure, cool air into the hive. By this simple and wonderful arrangement the bees are able to regulate the temperature to a nicety, for if it grows too warm, they have only to set more fanners to work, to expel the hot air. The temperature of the hive is a very important matter, for should it become too high the young ones would be suffocated, whilst if it dropped too low they would be starved to death.

PLATE XIX

Pollen gathers at Hive Door

The fanning is very hard work, and so, if we watch, we see that as a bee grows tired her place is taken by a fresh worker, and so the ventilating is constantly kept up.

During the hot nights of summer, in the busiest time, the hive is thronged with workers who have come home from the fields to shelter from the dew and cold of the night. The city then becomes very crowded and hot, and a large army of bees must be kept at work ventilating. If, on such a night, we were to steal down to the hive with a lighted candle and place it a few inches from the door, the draught caused by the fanners would be quite strong enough to blow out the flame!

CHAPTER XXIII
THE GUARD BEES

IF we watch for a short time at the city gates, we shall very likely see two bees apparently fighting desperately. If we look closely we may see that one of the bees has hold of the other by the wing, and is dragging it away from the door. To and fro the fight rages, and the bee which is held struggles fiercely, but without avail, for the other has her in a firm grip. The captive bee is really a robber, which has been caught whilst trying to slip into the hive to steal honey. It may be that the robber is from another hive, or perhaps is a wild bee, for there are communities of bees which are really like pirates. They have their homes in some hollow tree, and live either by robbing other cities, or by waylaying workers on their return from the fields, and taking from them the honey which they have so laboriously gathered. The bees, therefore, have found it very necessary that there should be a guard at the gates of their cities, and there are always some soldier bees on sentry-go.

To us, no doubt, one bee looks very much like another, and it is a mystery how the guards are able to recognise a strange bee. It is probable that the sense of smell has a great deal to do with this, for it is thought that all the bees of one hive smell alike, but differently from those of another hive, and that by this means the guards may detect a robber. A strange bee is never allowed to cross the threshold unless it is perhaps in the busy season, when the bees are "working overtime" as we might say, straining every nerve and muscle to gather in as much honey as they can before the summer goes and the flowers die. Then if a stranger comes to the hive, with her honey-sac full of the precious fluid, she may be allowed to pass in. Wasps often try to gain an entrance, as also do many other insects of one sort or another. If we watch the door for quite a short time in summer, it is pretty certain that we shall see several struggles. Sometimes it takes two or even three bees to expel the intruder.

On one occasion I witnessed a fight which lasted well over half-an-hour between a robber bee and a guard bee. They rolled over and over on the board, this way and that, each trying to get the better of the other. At last they fell on to the ground below, but even then they did not stop the fight, and the struggle continued on the grass.

Eventually the guard bee won the day, and by what appeared to be a final effort, she managed to pierce the abdomen of the robber bee with her sting. Instantly the robber bee was killed, and the brave little soldier bee returned to the hive in triumph.

It is not easy for one bee to sting another, for the abdomen and thorax are so hard that it can only be done through one of the rings of the abdomen, where the skin is thin.

CHAPTER XXIV
WORKERS IN THE CITY

BESIDES the fanners, the foragers, and the guards, there are other classes of bees at work in the hive. There are, for instance, the scavengers and cleaners-up, whose duty it is to keep the city and the combs spotlessly clean. Little twigs, dead leaves, and bits of gravel are all removed by these bees. Sometimes a mouse or a snail enters the hive, and then indeed there is great excitement. Imagine a great elephant-like creature, thirty or forty feet high, with a tail thirty feet long, to come walking into one of our cities, and you will have some idea what it seems like to the bees when a mouse is foolish enough to poke its head into the hive! But the bees are not frightened; the guards are promptly called out, and the poor mouse is soon put to death by hundreds of stings. Having made sure that the intruder is quite dead, the bees leave his body to the scavengers, who are confronted with the problem of disposing of it. If it were left it would cause disease and pestilence throughout the city, and it is too big and heavy for them to move. It is true that they might bite it into tiny pieces and thus carry it outside the hive, but this would take too much of the bees' valuable time. A better plan is thought of, and the body is soon covered over with a thin coating of wax. It is thus embalmed in a beautiful white tomb, which is made perfectly air-tight. If the tomb is near to the door, and interferes with the passing in and out of the workers, tunnels are cut through it. Sometimes when we look inside a hive, we may see two or three of these little mounds of wax, and we may be sure that each one is the grave of some intruder who had no right to be there.

Then there are the undertakers, who have a grim duty to perform. They carry away the bodies of workers who may have died within the hive, and in winter they have a busy time. It has been said, with what truth we do not know, that each hive has a burial-ground where the bodies of its workers are placed. It may be behind some bush in a corner of the garden, or perhaps down by the willows which fringe the banks of the stream. Whether this is so or not, it is certain that the undertakers carry the bodies of the dead bees away from the hive, so that they shall not pollute the pure air of the city and so cause disease. Now and then as we watch we may see one of these undertakers carrying what looks like the ghost of a bee! It is a bee in form, but its wings are folded, and its body is not a beautiful brown, but pearly white. This is a young bee, which has died before its birth, in the cell which has been both its cradle and its tomb. In winter, when it is too cold for the undertakers to journey far with their gruesome burdens, they will drop them just over the alighting-board, and so we sometimes see the ground near a hive strewn with dead bees, for many die during the colder months.

The water carriers are the bees who fly backwards and forwards between some neighbouring stream and the hive, supplying it with the water necessary to the workers. A hive should be placed near a stream or river, so that the bees may have as much water as they want, and they are helped in this if the stream be a shallow one in which there are little pebbles and rocks so that they can easily sip up the water. Another class of workers are the chemists, whose duty it is to place a tiny drop of acid, from their poison-bag, into each cell of honey, before it is finally sealed over. The acid supplied is chiefly what is called formic acid, and this is a very good preservative; it serves to keep the honey fresh and sweet until it is wanted.

You will remember that we said that it was actually good for us to be stung. This is because the formic acid which is pumped into the wound by the bee mixes with our blood, and prevents rheumatism. You will hardly ever find that a bee-keeper is troubled with this complaint.

CHAPTER XXV
THE COMB BUILDERS

IN order to trace the history of a hive, and to learn about the round of work which goes on day by day, we will suppose that a swarm of bees has been placed in an empty hive. We shall then be able to follow them as they commence with the first necessary work of building the combs. Our later chapters will lead us through the whole cycle of hive life.

We have already seen how the frames are placed within the hive, but we have yet to learn how the combs are built in them. Before the builders can set to work, however, it is necessary that the wax, of which the combs are constructed, should be made.

When a swarm of bees first enters the empty hive, numbers of them climb to the roof, and fasten themselves, by means of their tiny claws, to points of vantage. Other bees then join them, each hooking its claws in the claws of another, and in this manner chains of living bees hang from the roof in festoons. As time goes on these chains become more numerous, until the hanging bees look like a large cluster, for the chains cross and intertwine. All the bees do not form themselves into chains, for guards are posted at the hive door, while others examine every corner of their new home. The scavengers have to clean the floor and carry away twigs or gravel, so that everything shall be perfectly tidy for the builders to start work.

Now commences that wonderful and mysterious process of wax forming, which is carried on in perfect silence by the cluster of hanging bees. You will remember that the abdomen of the worker is composed of six rings; underneath these are the eight wax-pockets. There are two in each ring except in the first and last. It is perhaps interesting to note that the queen and the drone have no wax-pockets because they do not take part in the making of wax. For a similar reason their legs are not furnished with wax-pincers, like those of the worker. As the bees hang from the roof of the hive, in solemn and impressive silence, tiny scales are to be seen protruding from the wax-pockets. They look almost like a letter which has been pushed half-way into the slot of a pillar-box. A wax-pocket produces one wax scale, and so the workers each make eight tiny pieces of wax. In order that wax may be made in this manner it is necessary for the bees to consume a large quantity of honey, 10 or 15 lbs. of which produces only 1 lb. of wax.

We have already seen that the hind leg of the worker is provided with a set of wax-pincers (see Plate X.), and when the tiny scale of wax has been formed, these pincers take hold of it and remove it from the pocket. By means of the front legs it is then passed to the mouth, and here the strong little jaws come in useful. In its present state the wax is hard and rough, and it must be made smooth and pliable. It is mixed with juices supplied by glands in the bee's mouth, and worked by the jaws until it is so soft that it can be moulded into any desired shape. Often, when wax is being made, the floor of the hive becomes covered with wax plates which have fallen from the cluster above. When the wax has been kneaded to the correct degree of softness, the worker will leave the cluster of hanging bees, and crawl to the highest part of the roof of the hive. This is the foundation-stone of the combs, for they are not built upwards from the ground as our houses are, but downwards from the roof.

PLATE XX

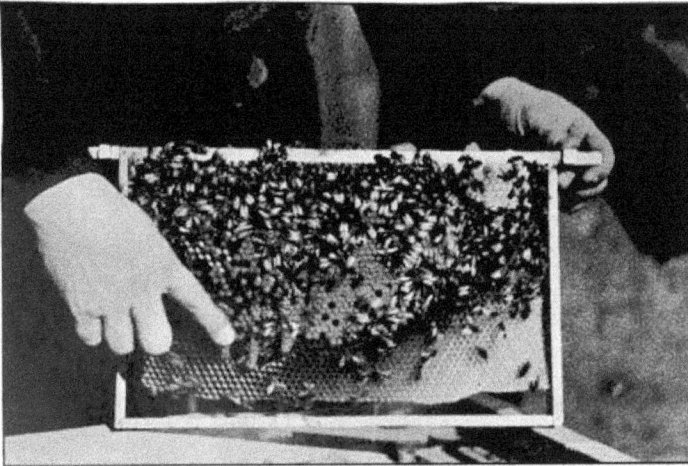

From a photograph by] [E. Hawks

Queen Cells on Comb

When the first plate of wax is in position, the little worker will take the other plates one by one from her wax-pockets, and knead them as she did the first. Each in turn will be placed on the foundation, and then the bee will again join the cluster. Immediately she disappears, however, her place will be taken by another, who goes through exactly the same process. She in turn will be followed by another, and so on, until a small piece of beautiful white wax hangs from the roof. At this stage it is time for the architects to plan out the position and shape of the first cells, which are to be sculptured out of the wax. If we watch, we may see one of these bees appear, and it is evident that she knows exactly what to do, and just what shape the first cell is to be. She moulds the unformed wax by means of her jaws, and very soon the outline of the cell is seen. It is hollowed out, and the wax removed in this process is carefully placed so as to form the walls. Meanwhile, another architect has been doing a similar thing on the opposite side of the piece of wax, for the cells are built back to back, as by this arrangement there is a saving of material. The wax-makers continue to add more and more wax, the sculptors go on with their work, and soon the form of the comb becomes apparent.

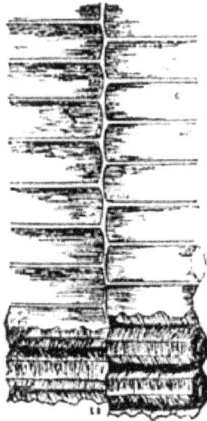

CRADLE CELLS.

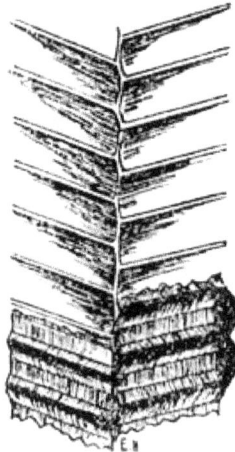

HONEY CELLS.

I suppose every one knows that bee cells are hexagonal, or six-sided. If they were made circular, you can easily understand that there would be a great deal of space and material wasted, for the spaces between the cells would need to be filled up. Then, again, if they were made diamond-shaped, there would still be places to fill in. It is true they might be made four-sided, but apart from the fact that such cells would not be strong enough, it is not possible for them to be made thus, for the angles would be too great for the bees to get their jaws into the corners. It has been found that six-sided cells are the strongest and the most economical, but how the bees found this out, too, is a mystery.

There are three kinds of bee cells: firstly the cradle cells, in which the young bees are reared. They are $\frac{1}{2}$ inch deep and $\frac{1}{5}$th inch in diameter. There will therefore be about twenty-eight in a square inch of comb, but as the drone is slightly larger than the worker, his cradle must be bigger. We find accordingly that the drone cells are $\frac{1}{4}$th inch in diameter, or about eighteen to the square inch.

Then there are the royal cells, which are altogether different. In them the young queens are reared, and in appearance they are something like acorn cups. In Plate XX. you see a picture of a frame of comb, taken from the hive with the bees still on it. The bee-man is pointing to two of these queen cells, and you will see that they hang downwards, in a place where the ordinary comb has been cut away to make room for them.

Lastly there are the honey cells, which are of the same size as the cradle cells, but instead of being built horizontal they are made sloping upwards. By constructing them in this way honey stored in them is prevented from running out over the combs.

The back of the cells, or the dividing wall between the two sets, is not flat, as we might imagine. If you look at the sketches you will see that the cells are fitted into one another so cleverly that the bottom of one cell forms half of the bottoms of two cells of the other side of the comb. All the cells of one sort, say for instance the honey cells, are made exactly the same size, and do not differ by the fraction of an inch. How the bees are able to measure the width when building them is a mystery. Perhaps the antennæ have some important part to play in this matter, but if so it has yet to be discovered. Another thing which is as curious as it is mysterious is how the sculptors on each side of the comb are able to fit in the cells so neatly that each one is in its right place with regard to the cells on the other side of the dividing wall. It is certain that the workers cannot see through the wall of wax, and yet the two lots of cells correspond exactly.

CHAPTER XXVI
THE LIFE OF THE BEE

ALL the time the cells are being built the queen wanders about the hive in a distracted fashion, because there are no cells ready for her to fill. Now that some are ready, however, her movements change. Surrounded by her councillors, or ladies-in-waiting, as we might call them, she clambers over the comb and selects a cell in which to lay the first egg. She very carefully examines the cell by placing her head in it and feeling the sides with her antennæ. Being satisfied that it is in a fit state to become the cradle of a young bee, she withdraws her head and then the egg is laid. All this time the ladies-in-waiting stand round, and in the season for egg-laying you may quickly pick out the queen by the circle of bees about her (see Plate III.). They guide her over the comb, feed and clean her; sometimes, too, we may see them stroking her very tenderly with their antennæ. After the first egg is deposited in the cell, the queen moves to the next, and so on all through the summer. During this time she lays day and night, and does not appear to sleep.

The eggs are little pearly-looking objects something like tiny rice grains, and each one is fastened to its cell by a drop of gummy liquid.

In the meantime the bees are at work building combs with all haste, for the queen is close on their heels, demanding more and more cells. She does not rest until the whole of the ten or twelve frames have been completely filled with cells and eggs. By this time the first eggs which were laid will have hatched out into young bees, who will leave their cradles to take part in the duties of the hive. These first cells will then be cleaned out by the scavengers, and the queen will lay more eggs in them. In this way the queen goes on all the summer, and as a matter of fact, if the hive be a prosperous one, she may lay as many as 3000 eggs each day! After the eggs have been laid the queen does not appear to take the slightest interest in what may become of them. On the other hand, the worker bees do, for they know that on these tiny little eggs depends the future of the hive.

In three or four days an egg will hatch into a tiny white grub, which the nurse bees immediately commence to feed. It is not fed upon honey, though, for that would be like feeding a baby on roast beef! The nurse bees have certain glands in their bodies by which they are able to turn honey into a kind of bee-milk, and this is called "chyle food." For three days the little grub is carefully fed upon this preparation, and then it is given "modified chyle food," as it is called, which is also bee-milk, but richer than before.

During these few days the grub casts its skin and grows very quickly, until on the fifth day it turns into a chrysalis, just as a caterpillar does before becoming a butterfly. The bee-grub spins a soft silken cocoon, and the sculptor bees come along and seal over the mouth of the cell with a cover, which admits air so that the grub may breathe.

The grub then commences what is called its *metamorphosis*—a Greek word meaning "a change of form"—and a wonderful change it is. In sixteen days from the time that the cell was closed up, the fat little grub turns into a perfect worker, just like a caterpillar changes into a butterfly. The young bee is now ready to emerge from her cell, and the porous capping is the only barrier. The little prisoner, however, finds that she has a sharp pair of jaws and so begins to bite the capping. Slowly it is all snipped away, and we see a tiny hole appear, which grows larger and larger. In a few moments out comes one of the antennæ, and waves about as though to explore the world beyond the cell. It seems to give a good report to the little bee, for the biting of the cap is redoubled, and before long, assisted perhaps by some of the nurse bees, the youngster slowly emerges. She is, however, very pale and weak as yet, and so the nurse bees commence to clean and feed her. She soon gains sufficient strength to take an interest in what is going on around, and we may imagine that she is somewhat surprised to find how busy is the city into which she has stepped—every one rushing here, there, and all over, none seeming to take any notice of the young bee, and everybody apparently having something to do, and to be in a great hurry to do it!

A fortunate insect is the little bee, none the less; for she has no need to attend school or to have any lessons. She knows all that she need know as soon as she is born. In a few hours' time, for instance, she will be feeding grubs, just as she was fed by other bees some days before. She will know all about the city, the duties which she has to perform, and the respect which she must pay to the queen, her mother. After perhaps a fortnight or so of nurses' work she will join the ranks of the foragers, and seek the nectar of the sweet-scented flowers.

PLATE XXI

From a photograph by] [E. Hawks

Queen Cells

This, then, is the history of the birth of a worker bee, of which a prosperous hive may contain anything from 30,000 to 60,000. The history of the birth of a drone is practically the same, except that in his case it takes twenty-five days for the egg to change into the complete insect.

CHAPTER XXVII
THE STORY OF THE QUEEN

AMONG most nations it is customary for the kingship to be handed down from father to son, but no such rule exists in the bee-city. Although we call one of the bees the Queen, she is not really a queen in the ordinary sense of the word. She does not rule the hive, nor can she command the bees to do this thing or that, and a far better name for her would be the Mother bee.

Up to the seventeenth century it was thought that a hive was ruled over by a king-bee, and it was not known that this large bee was the mother of all the other bees, and yet this is so, as we have already seen. Whether or not a queen shall be born depends on the wish of the workers, and it is surprising to find that a queen is developed from an ordinary egg, which, if it were not subjected to certain different processes, would turn into a worker bee.

PLATE XXII

From a photograph by] [E. Hawks

An Empty Queen Cell

When the bees desire that a queen shall be born, the builders and sculptors are first consulted. They set to work to make three or four queen cells, or, as we might call them, royal cradles; in one of them the future queen will be reared. We have already seen that queen cells are different from the ordinary cells, and that for their accommodation a part of the comb is cut away. This gives better ventilation, and the royal cells hang downwards from the comb as seen in Plate XXI. The nurse bees now place in the first an egg from one of the worker cells, but this egg must not be more than three days old, otherwise a queen would not be produced, no matter what efforts the bees might make. Eggs are placed in the other cells at intervals of three days. On the fourth day the first egg hatches into a grub, just as it did in the case of the worker bee, whose career it resembles up to this stage. But now the nurse bees, instead of feeding it upon chyle food, commence to supply it with "royal jelly" as it is called. This is a very rich form of food, and is only given to those grubs which it is intended shall become queens. The nurse bees continue to pay special attention to the little grub, and give it as much of the royal jelly as it can take. This goes on until the ninth day, when the grub spins a cocoon and the cell is closed up. On the sixteenth day from the time the egg was laid the young princess will be ready to leave her cell; she will then commence to gnaw the floor in order that she may get out. In Plate XXII. there is shown an empty queen cell, the floor of which has been cut away in this manner.

Thus we see that the making of the queen rests entirely with the workers themselves, and depends simply on an egg being placed in a certain kind of cell, and having special food and plenty of ventilation. After the queen has been hatched, the royal cell is cut away, and its place filled with honey cells. The wax of the cell is not wasted, but used in the construction of new comb.

CHAPTER XXVIII
THE POLLEN GATHERERS

LET us now follow one of the pollen-gathering bees on her quest of bee-flour, which is so necessary for feeding the inhabitants of the hive. Having first taken a careful survey of the position of the hive and its surroundings, our little worker flies off at top speed to the hillside or the orchards where, it may be, the fruit trees are in full bloom. On her way, perhaps, she will decide what kind of pollen is to be gathered, for different kinds of pollen are kept quite separate, just as our own flours are separated. It remains a mystery why bees should keep the different pollens apart, as it seems to us that it would not matter much if they were mixed, but no doubt the bees know better than we do. Although buttercups may be scarce, and though the hedges are laden with hawthorn blossom, yet if the gathering bee has started to collect buttercup pollen, she will pass by the hawthorn and search diligently for buttercups in the adjoining meadows.

PLATE XXIII

Storing the pollen in cells

Arriving at the flower, the little worker alights and moves about it, so that very soon her hairy body becomes covered with pollen, as shown in the *frontispiece*. Although she was a brown bee when she alighted on the flower, now she is all golden yellow, and looks like a dusty miller. It is here that the brushes and combs with which the legs are furnished come in useful, and after two or three flowers have been visited, we may see her brushing down her body, and combing the pollen grains out of the hairs in which they are entangled. The collected pollen is then moistened with a tiny drop of honey, and kneaded into little round pellets, which are placed in the pollen baskets. This being done, the bee flies on and on, visiting other flowers, until her baskets are quite full. Sometimes the bee gathers more than can be carried in her baskets, so she returns to the hive with her body smothered in gaily-coloured pollen.

Though her wings are strong, yet the load of pollen is heavy, and all her strength is needed to reach the hive in safety. It may be that she is almost exhausted before she can alight on the board at the city gates. So she will settle on a leaf or some flower, like a ship coming to anchor, in the harbour of the garden, and here for a few seconds she will rest, to gain fresh strength for the final flight. Some of the bees seem to act as inspectors, or general helpers as it were, always on the lookout to do somebody a good turn or to lend a helping hand wherever it may be required. And now, as the pollen gatherer makes a final flight to the board, these bees come forward and help her to drag her load safely within the city. Once inside the door, the worker makes straight for the cells which might be called the flour bins, for here the pollen is stored. A picture of them is to be seen in Plate XXIII., and you will notice that the different kinds of pollen are still kept separate. Arrived here, the gatherer levers the pellets out of the baskets by means of the spurs on each of her middle legs. These act as little crowbars, and the pollen is then placed in the cells. If it is not intended for immediate use, some of the house bees will cover it over with a layer of honey, for it would not keep if left exposed to the air. We should imagine that the pollen gatherer would now take a rest, or at any rate some refreshment. This, however, is not the case, for no sooner has she got rid of her load than she darts towards the door, and before we have time to follow her she is off to the fields again for another load. From morning to night she continues to travel backwards and forwards between the flowers and the hive. Is it any wonder, then, that at the end of a few weeks' time the brave little worker will have completely worn away her wings, and will lie down and die?

When watching the alighting-board, you will remember that we remarked on the pollen gatherers entering the hive, each with the little baskets filled with bright-coloured pollen; from the colour of the pollen we may tell from what flowers the bees have brought it. The deep golden-brown comes from the gorse bloom, away on the hill; the snow-white from the hawthorn, and the vivid yellow from the buttercup, or perhaps the dandelion. The pale green is from the gooseberry bushes, whilst the pollen of the charlock is golden and clover pollen is russet-brown. Sometimes, when the poppies are growing among the corn, the little gatherers will return with loads of jet-black pollen, while the orchards give many delicate hues, the most beautiful of which is the light yellow from the apple blossom. On rare occasions, we may see a worker come laden with pollen of deep crimson, but the source of this wonderfully coloured stuff is a mystery, for we do not know from what flower it is obtained.

CHAPTER XXIX
THE VARNISH MAKERS

SOME people think that bees gather only honey and pollen, but there is another substance which they collect, and this is called "propolis." The poplar and pine trees have, as perhaps you know, a resinous kind of matter covering their new shoots, whilst the horse-chestnut protects its leaf buds with a similar sticky substance. This the bees gather, and they draw it off the trees in thin strings, just as sometimes you see children playing with a piece of sticky toffee, by pulling it into two pieces. The bees then roll these strings into balls, and pack them in their pollen baskets, and return to the hive. The other bees help to unload as soon as the gatherers arrive, for the sticky substance soon hardens, and must therefore be got out of the pollen baskets as quickly as possible, and for the same reason it must be used at once. The bees then knead it with their jaws and mix with it some liquid from their mouths, until it is quite soft and pliable.

With this preparation, which is really like varnish, the bees coat the whole of the inside walls if the hive is a new one. Should there be any cracks in the walls or floor, they are carefully filled up to keep out the cold and damp. Then again the propolis, in a stronger form, is used for fastening the combs to the frames, and for any other objects which the little engineers may think need firmly fixing. When we open a hive we find that the felts, which cover the combs and keep them warm, are firmly fastened down to the frames, and sometimes we have to use considerable force to get them off. The frames holding the combs are fastened into position, too, with propolis, and a mixture of this substance and wax is used to cover over the bodies of any intruders who have entered the hive and have been stung to death. The combs containing sealed cells of honey are subjected to a coating of very thin propolis to keep them sweet and clean. Plate XXIV. is a photograph of a frame of comb just removed from the hive. Towards the top you will see bees busy capping the honey cells, and others are varnishing them over with propolis. The cells inside the white lines are pollen cells, and you may see pollen-pellets in them.

PLATE XXIV

From a photograph by] [E. Hawks

A Frame of Comb, showing Bees at Work storing Honey and Pollen

CHAPTER XXX
THE NECTAR GATHERERS

IN this chapter I propose to relate to you the day's work of a nectar gatherer, or forager. These are perhaps the most important workers in the hive. If you look it up in your dictionary you will find that nectar is described as being "any pleasant liquid." I want you to understand that the bees do not actually gather what we call honey. What the bees gather, and what the flowers secrete, is nectar, which is a thin watery liquid, containing among other things a large proportion of cane sugar.

Arrived at the meadow the forager alights on the first suitable flower she comes to, and dips her tongue down to the nectaries. Even the tiniest droplet of nectar can be collected by means of the spoon at the tip of the proboscis. She visits flower after flower until her honey-sac is filled, and then she sets out on the return journey to the hive. Whilst she flies a change takes place within the honey-sac. First of all the nectar is strained, to separate the pollen, and in the manner we have already seen. Then some juices are added to it which are supplied by glands in the bee's body. The cane sugar is changed into another form, called grape sugar.

Cane sugar is not good for either us or animals to eat, but on the other hand grape sugar is beneficial. You will know that we cannot derive any nourishment from our food until it has been acted upon by the saliva of the mouth and by certain juices in the stomach. The food is then said to be digested. Practically the same change is carried out in the bee's body, the nectar being converted into honey. In her case, however, the change is not made only upon the food she consumes herself, but also on that contained in the honey-sac. Many people think that the honey they eat is just in the same state as it is in the nectaries of the flowers from which it has been gathered, but now you will know that this is not so. The reason that honey is good for us is that it has already been partly digested by the bees, and therefore our stomach is saved a certain amount of work.

Our bee has now arrived at the hive, and as she passes the guard bees she is recognised as being one of themselves, and her entry to the hive is not delayed. The guards may salute her as she passes, with a wave of their antennæ, and she hurries off to the storehouses. Here the warehouse bees are kept busy storing away the honey brought in by the foragers, and to one of these bees our little friend hands over her load. At least she does not "hand" it over, but passes it from her tongue to that of the other bee, who in turn swallows it. This bee then climbs to the cell she is filling, and placing her tongue therein, empties the honey into it. No sooner has the forager been relieved of her load than she makes her way to the hive door, pushing and struggling, butting with her head here, or crawling over her sisters there, until she at last forces her way through the crowd and flies off to gather further supplies. A bee that is one day gathering nectar will probably collect pollen the next day, and *vice versa*. By this arrangement the organs which change the nectar into honey are given a rest.

We cannot tell how bees are able to find their way home to the hive so cleverly. They may fly two, three, four, or even more miles away to the flowers, but they are always able to return. If a forager bee is imprisoned in a box, and carried a couple of miles away and released, she will reach the hive long before we could; in this respect you will see that bees are something like homing pigeons.

CHAPTER XXXI
THE WINTER SLEEP

DURING the summer the bees work only with the idea of storing away sufficient honey to last them during the dark days of winter, when there are no flowers. In the tropics, where perpetual summer reigns, the bees live as it were from hand to mouth, and do not store nearly so much honey as those bees which live in climes where the summer is followed by a long winter.

When autumn comes, and the flowers vanish, the bees gather round the queen on the combs of the hive; we see some of them in Plate XXV. The builders block up the doorway with wax until only a narrow passage is left, just large enough to allow them to travel in and out. This is done to keep out the cold of winter, for then it is necessary for the temperature inside the hive to be as high as possible.

In this cluster the bees pass the winter in a kind of sleep. They eat the honey which they have stored, and wait for the arrival of spring. The outermost bees of the cluster are of course the coldest, and so that each may take a turn at being on the outside, they constantly change places. They only leave the hive on a few occasions during this time, and then it is to take a short flight for exercise.

PLATE XXV

From a photograph by] [E. Hawks

Bees clustering in Winter

When the bright sunshine comes, and the crocuses tell of the coming of spring, the bees begin to bestir themselves. Sometimes when it has been snowing, and the snow is lying on the ground, the bees are deceived by the glare into thinking that spring has come; they fly out to look for flowers, but many of them are killed by the cold. When spring is actually at hand, however, the pollen gatherers are despatched to the crocuses and other early flowers. They come back laden with pollen, and as soon as the queen bee sees this she commences to lay. These eggs will develop into the bees which will carry on the work of the hive during the summer. The bees which have slept through the winter only live long enough to look after these eggs, and to bring the young bees safely into the world.

It is interesting to note that the amount of nectar and pollen gathered will, to a certain extent, regulate the number of eggs that the queen will lay. If food is scarce she will not lay many, for if she did a great number would have no food and all would die of starvation. If, on the other hand, honey and pollen are abundant, hundreds or even thousands of eggs will be laid in a single day. The number is increased, too, as spring merges into summer, and for a fortnight or three weeks in May or June, the hive is at its busiest. During this period the fields are white with clover, and the flowers are at their best. This time is known as the honey flow, and if the hive be a prosperous one, the honey does literally flow into the combs.

CHAPTER XXXII
THE SWARM

IT is not known exactly why bees swarm, and it has been said that it is because the hive becomes overpopulated. When the hive becomes crowded early in the summer, the bees build queen cells, and in them royal princesses are reared, as we have already seen. When the time approaches for them to leave the cells, the old queen begins to get very excited, for she seems to know that a rival is about to be born. She would like to rush to the cells and put the young princesses to death, and indeed she would do this, were not the cells guarded by the other bees, who anticipate trouble with the old queen. So, though she may make the attempt, after being repulsed time after time she will give up, and adopt another procedure. She seems to realise that her rule in the hive is at an end, and so she determines to leave it on the first fine day, with as many of the other bees as will accompany her, and to fly to pastures new. All is then commotion with the bees that will go with her, and they seem to eagerly await the signal to be off. No one knows how it is decided which bees shall go, or which shall remain, for old or young, builders or foragers, may go or stay. All who are going, however, take in supplies of honey, and when the appointed time has arrived the swarm issues from the door of the hive in a thick black stream. The old queen will be among them, and they generally fly to some tree close at hand. A suitable spot is chosen, on one of the branches perhaps, and the leading bees settle there. These are quickly joined by the others, so that in a few seconds the cluster is as large as an orange. It grows larger and larger, until after a few minutes from the time the bees left the hive in a mad throng, they will all be quietly hanging in a pear-shaped mass like those in Plate XXVI.

PLATE **XXVI**

From a photograph by] [W. Dixon

A Swarm

A swarm is a wonderful sight, for the bees are almost perfectly still, and hang in a glistening mass, clinging to one another by their tiny hooked claws. Sometimes the leading bees of a swarm choose queer places in which to cluster: one lot, for instance, swarmed on to the beard of a gardener, whilst another found a resting-place on the neck of a horse which was standing under some trees!

As soon as the bees have swarmed on the branch, or wherever they may have settled, scouts are sent out to look for a suitable place for the new home. They return with news of some spot which they think would serve the purpose. This scout thinks that the hollow tree she has found would be best, but another says that a little cave in the rocks would be better. Meanwhile more scouts are despatched, and when all the different proposals have been considered, and all possible places discussed, it is finally settled where the future home shall be. Headed by the scouts, who now act as guides, the swarm then takes to flight once more, and will not stop until it reaches the chosen spot. Wherever or whatever it may be matters not, for the bees will have to commence at the very beginning of the cycle of home life, and as soon as they are all inside the new home the wax-makers will climb to the highest points, hang in chains, and begin to make the wax for the combs, exactly as we saw in a previous chapter.

PLATE XXVII

The Bees in their New Home

CHAPTER XXXIII
TAKING THE SWARM

BEE-KEEPERS watch for the issuing of the swarm, and when it occurs they get ready to take it, so as to fill another hive with bees. Having found where the bees are hanging, an empty hive is brought to the spot and placed under the branch. The bees are then shaken into it, or they may be even gathered in handfuls, or with a spoon, and placed in the new hive (see Plate XXVII.). At the time of swarming bees are practically harmless, for they have taken so much honey that they do not feel disposed to sting. The old straw skeps are often used for taking a swarm, for they may be more conveniently handled than the larger and heavier wooden hives. The hive which is to be their permanent home is placed close at hand too, with a clean white cloth on a board leading to the door in front of it. After the bees have been shaken into the skep they are emptied on to the cloth, and at once commence to walk into their new home (Plate XXVIII.). There are thousands upon thousands of bees in a swarm, and pictures of them going into new hives are shown on Plates XXIX. and XXX. In order to show how harmless the bees are at swarming time, the bee-man in the first picture has taken a handful of them, as we may see.

There is an old rhyme which says:

"A swarm in May, worth a load of hay,

A swarm in June, worth a silver spoon,"

and the bee-keeper is pleased should his bees swarm in May, for then he will be able to put them in a new hive, and they will gather a good supply of honey before the summer is over. Should the swarm take place a month or two later, however, the bees do not settle down in time to gather sufficient honey for the winter, and they cause the bee-keeper trouble, for he has to feed them with syrup.

After a swarm, the bees seem to forget all about their old life and companions, for the hive containing the swarm may be placed quite close to the old hive without either set of bees taking the slightest notice of the other.

If a bee-keeper is not at hand to take the swarm, the bees will probably make their home in some hollow tree. They will commence to build combs, and young bees will be reared and honey stored just as in a hive.

PLATE XXVIII

From a photograph by] [E. Hawks

Thousands of Bees walking into their New Home

CHAPTER XXXIV
THE OLD HIVE AFTER A SWARM

AFTER the old queen has left with the swarm, the bees have to decide what to do about a new queen, and the eldest princess is, as we have already seen, clamouring to be set free from her cell. Although she gnaws away at the floor of her cell the bees keep her a prisoner, by piling more wax on the outside of the cell. She is kept thus until the old queen has got away with the swarm, otherwise there would be a terrible fight between the rival queens.

However, the bees now decide to set the young princess at liberty, and two courses are open to them. If the hive has got what is called the "swarming fever," the princess will lead a second swarm, for she knows that in a few days another princess will be born. This second swarm is called "the cast," and unlike the first, flies away at once, no matter what the weather may be, for there is no time to be lost. The cast does not settle near the hive as the first swarm does, but flies quite away, and is generally lost to the bee-keeper.

If, as is generally the case, the hive has not got swarming fever, the bees adopt the princess as their queen. As soon as this course is decided upon, the bees allow her to visit the cells containing her rivals, and with savage anger she inserts her sting in each cell and puts them to death.

During the next few days she wanders about the hive in a restless fashion, constantly going to the door. After a while she leaves the hive and flies high up into the air. She is not allowed to go alone, however, but is followed by numbers of drones. In about an hour's time she returns, and the bees know that she is now mated and will remain quietly in the hive. The hive then resumes its ordinary life, and the young queen commences to enter upon her new duties. The queen cells are no longer required, and so they are cut down; the builders set to work to erect nursery cells in their place, for every available inch of room will be required by the young queen for laying eggs.

PLATE XXIX

Bees going into a Skep

CHAPTER XXXV
THE MASSACRE OF THE DRONES

WE have already seen that the drones do no work in the hive, nor do they gather nectar or pollen. They live a life of ease, feeding upon the honey gathered by the workers, and it has been said that each drone eats as much food as can be provided by four workers. You may understand from this that the drones would quickly eat up the provisions which have been gathered for the winter. The workers know this too, and when the summer begins to go and the flowers to fade, the drones will meet their fate. They are always under the power of the workers, for besides eating honey, they are given chyle food, and were the workers not to give them this, at the end of three days the drones would die, even though there was plenty of honey around them.

It is not by starvation that the drones die, however, for they are massacred by the bees. Some time about August, perhaps, when the bees find that the honey is not coming in as fast as it used to, the step will be taken, for they have now to think about the winter months which are close at hand. If there are any drone cells in the hive with eggs or grubs in them, the workers tear them open, the young drones are dragged out, and their bodies thrown out of the door of the hive. Although the other drones may see these proceedings they take no heed of them, but continue to live their lazy life, and to eat their fill of honey. But in a few days the signal for the massacre is given, and the workers commence to put them to death. Throughout the hive may be seen the workers chasing the drones over the combs which, but a few days before, supplied them with honey. The drones have no sting, nor any means of defence, so that they are absolutely at the mercy of their pursuers. The bee-city is alive with the terrible cries of the victims, and as the workers catch the drones they commence to bite off their wings. Sometimes, too, they will even gnaw off the legs or the antennæ or cut through the drone's slender waist, their one idea being to disable him. Some of the drones perhaps are able to escape from the hive, and may seek refuge in flight, but after a few hours they are back again. They cannot live without food, and as they have never done any work, they do not know how to gather it. When they return, the guard at the gate, which is always doubled at this time, savagely fall upon and kill them. Some do not return to the hive, but these speedily perish of cold when the night air comes on. The bees never sting the drones in the struggle, for the sting, being barbed, would soon be pulled out by the roots were it once inserted in the drone's body. The bodies of those that have been killed are carried out of the hive by the undertakers, and a busy time they have, as you may imagine.

PLATE XXX

Half-an-hour after Plate XXIX

CHAPTER XXXVI
HONEY

IN the old days the people did not know where honey came from. Pliny, the great Roman writer, says that it came from the air, and that the stars helped to make it. He tells us, too, that it was much better at the time of the rising of the bright star Sirius, and goes on to say what a pity it is that it is mixed with "the juices of the flowers," for, little suspecting that they are really the nectar itself, he actually thought they spoilt its essence! Others supposed that honey gathered whilst Jupiter and Venus were in the sky with Sirius was able to effect miracles, such as curing diseases and actually restoring the dead to life! How curious and interesting are these old beliefs, and yet how silly they seem to us. We know that honey is really "juices of the flowers," which have undergone a wonderful change in the bees' stomachs; and that, although it is very pleasant to the taste, it is not able to restore the dead to life, or to work any other miracles of a like nature!

There are many different qualities of honey, each depending on the flowers from which the nectar is gathered. There is, for instance, the beautiful almond-flavoured honey from the apple blossom or the dark and strong heather honey. But the honey which is perhaps the most common and beautiful is that from clover. The white clover blooms for about three weeks and then indeed are the bees busy. Red clover is of little use, the florets being too long for the bees to reach the nectar. It is true that this might be obtained by the bee biting through the base of each one, but when red clover is in bloom the white is also to be had, and so of the two the bees naturally prefer the white, where their tongues can easily gather up the tiny drops of fluid. Later on, perhaps, when the white clover is done, there will be a second crop of red, and the bees are then glad to visit it, for the florets of the second crop are shorter than those of the first. Clover honey is light amber colour, and as clear as crystal. A bee-keeper can tell by the taste of any honey from what flowers it has come, and perhaps, too, from which part of the country.

After the bees have filled up their combs with honey, the bee-keeper puts some smaller frames in the second chamber of the hive. These are called sections, and as a rule they measure $4\frac{1}{4}$ in. $\times$ $4\frac{1}{4}$ in. If honey is still plentiful the bees will then build combs in these sections, and fill them with it, and so when this has been done the bee-keeper may take away the sections, and it is in this manner that honey is taken from the bees.

Each section contains about 1 lb. of honey, and you may often see them for sale, at about one shilling each. Some people prefer honey when it has been extracted from the sections and put into glass jars; myself, I think it is far nicer to eat it from the comb. An average hive will give about 30 or 40 lbs. of honey a season, but you can easily imagine that a great deal depends upon the weather. The situation of the hive counts, too, for hives in the south of England give more honey than do those farther north. This is because the flowers in that part are much finer and yield more nectar, and also because the climate is warmer.

PLATE XXXI

Bees on White Clover

CHAPTER XXXVII
MODERN BEE-KEEPING

WE have already seen that straw hives were formerly used to keep bees in. They had many disadvantages, and perhaps the greatest was that sections could not be put on to collect the extra honey. The only way in which it could be obtained was to kill the bees and to take the honey they had stored for themselves. The bees were generally suffocated by the fumes of burning sulphur, and so you will see that besides being inconvenient this method of bee-keeping was also very cruel. The hives with the greatest number of bees were the healthiest, and they were selected for treatment in this manner, for they had more honey stored away than the weaker ones. In this way all the best bees were killed off, and those that we have at the present time are descended from poor ancestors. It will be many years before they have been brought back to their former state of excellence.

After the bees had been suffocated, the old bee-keepers took out the combs. These were not built in frames as are those of the present day, but were just made inside the skep in any way the bees liked. The honey was then extracted from them, but it was of very poor quality, for pieces of broken comb, pollen, and even dead grubs, or parts of the bees themselves, were mixed up with it. How different this is from the beautifully clear honey obtained by the modern methods.

PLATE XXXII

From a photograph by] [E. Hawks

Sealing over the Honey Cells

After the cells have been filled with honey the bees leave them uncovered for a little time, so that the water in the honey may evaporate. The honey then ripens, and the chemist bees place a tiny drop of formic acid in each cell. When all is ready, the cells are sealed over, and in Plate XXXII. we may see the bees at work doing this. You will be interested to know that the English bees do not quite fill the cells, and so the colour of the honey does not show. Foreign bees, however, fill the cells quite to the brim, which gives the comb a dark and dirty appearance.

Nowadays the straw skeps are very seldom seen, for their place is taken by the wooden hives we have already considered. The frames containing the combs are all of the same size, so that they may be transferred from one hive to another. For instance, should a certain hive have collected a large quantity of honey for winter use, and another hive not have sufficient, the bee-keeper may take one or two frames of this honeycomb from the rich hive and put it into the poor one, and in this way both lots of bees will live throughout the winter. In many other ways the frame hives are useful, besides being much more healthy. The bees need not be killed in order to get the honey, as was necessary with the skeps, for a puff or two of smoke is all that is required, and while they are frightened we may remove the sections.

You will understand that the sooner the queen sees pollen coming into the hive in the early spring, the sooner will she commence laying eggs. The sooner the eggs are laid, the more bees will there be ready for the summer flowers. So the bee-keeper sprinkles pea-flour in a box of shavings near the hive in the early days of spring. The bees soon find the flour, and, thinking it is pollen, they commence to carry it into the hive. When the queen sees it coming in she is deceived, and thinks summer is at hand; so she commences to lay eggs. This gives the hive a start, so that when spring really comes, there are large numbers of bees ready to gather honey from the early flowers.

We have already mentioned that a great quantity of honey has to be consumed before wax can be made, and this is a serious loss to the bee-keeper, for it not only reduces the stores, but also wastes valuable time as well. So the bees are now provided with a thin sheet of wax, a piece of which hangs downward in each frame. On it is stamped the exact design of the cells, so that not only is material provided for the bees, but the architects are saved the trouble of having to map out where each cell shall be. A piece of this "foundation," as it is called, is shown in Plate XXXIII. The bees readily take to it, and as soon as the work of building is to commence they knead the wax and draw it out from the foundation, until it is a complete cell. In this way a great deal of time is saved.

PLATE XXXIII

From a photograph by] [E. Hawks

Foundation, showing the Pattern for Cells

CHAPTER XXXVIII
THE BEES' ENEMIES

BEES have many enemies, apart from robber bees, who try to steal their honey. In winter-time, when pressed by hunger, certain birds come to a bee-hive and commence tapping on the alighting-board. Of course some of the bees come to the door to see what is the matter, and no sooner do they appear on the threshold than the sharp little birds grab them in their beaks, and so make a meal. Birds often catch the bees as they are gathering nectar in the fields, and no one knows how many perish in this way.

Then there is the death's-head moth, as it is called. You no doubt know that this is an insect which bears on its back markings like a skull, and hence its name. It sometimes enters a hive and makes a chirping noise. It is supposed that this fascinates the bees, and the moth is therefore able to take whatever it wants in the way of food.

Bees have fleas too, and though they are not very formidable enemies, they are a nuisance. A picture of one of these tiny mites is found in Plate XXXIV.

The worst enemies of the bees are diseases, of which there are several kinds. The most dreaded are dysentery and what is called the "Isle of Wight" disease. Many of our soldiers died of dysentery in the South African War, caused through their drinking bad water, and it is the same kind of illness which attacks the bees. The Isle of Wight disease is as peculiar as it is mysterious. It resembles the dreaded sleeping sickness from which natives of Africa suffer, and of which we have heard so much these last few years. The bees seem to lose all power of flying, and in a few days whole hives may die. It is called the Isle of Wight disease because it first appeared in that island a few years ago.

PLATE **XXXIV**

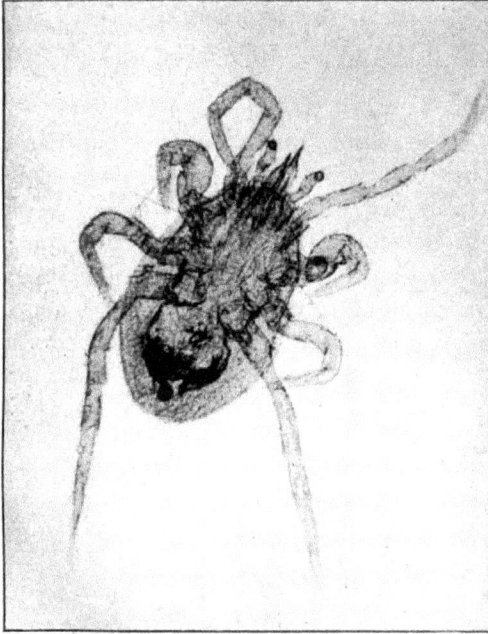

From a photo-micrograph by] [E. Hawks

Parasite of Bee

CHAPTER XXXIX
POWERS OF COMMUNICATION

BEES have not, so far as we can tell, any system of language such as we have, but it is quite certain that they are able to communicate with one another. Not only can they communicate simple facts, but they actually can, in their way, talk or tell each other things. How this is accomplished without any voice we are not able to say, but it is certain that in this connection the wondrous antennæ play a most important part. If you watch bees on the board in front of the hive, you will see them sometimes march up to one another and gently cross the antennæ, as two duellists cross their swords before a fight. For a fraction of a second one seems to lightly tap the antennæ of the other, and it is evident that some communication is passing between them. It may be some important piece of news, or perhaps it is just some hive gossip, of interest to both the little insects. Who can tell?

An experiment which I have often tried with bees, to show that there is the power of communication, is to put a few drops of honey on a saucer, which must then be placed at some distance from the hive, or there would soon be a crowd of bees round it. Next, a bee is entrapped and placed on the honey. She will commence to sip it up, and as soon as she has taken as much as she can carry will fly to the hive. When next she comes back for honey she will probably be accompanied by a friend; on the third or fourth visit, if the honey still lasts, several more bees will also visit it, and all will be busy carrying it to the hive. I should tell you, however, that it does not always happen that the first bee will bring friends. I have tried the experiment many times, and have come to the conclusion that there is no doubt the first bee does often tell other bees of her find, and that they come to help her to gather in the treasure. In this regard a still further experiment may be of interest. Many of you no doubt have seen that beautiful fairy play called *The Blue Bird*. This was written by an author called Maurice Maeterlinck, who has also written a very interesting book, *The Life of the Bee*. Mr. Maeterlinck has suggested for this experiment that honey should be placed on a plate or saucer some distance from the hive, as in the other case. Then a bee should be put to the honey and allowed to take in a supply. While she is feeding she will be so deeply interested that we are easily able to mark her by painting a tiny spot of colour upon her back. Now away flies the bee to the hive, and hands over the honey to the house bees. She will then leave the hive and fly back to the plate for more honey. She must be trapped as she leaves the hive, and kept in a little box. Now if bees have the power of communicating, we might expect that the marked bee would have told some of the other workers of her find. So far so good, but what we wish to know from this experiment is whether or not the marked bee was able to tell the other bees where to find the honey, or whether she only said to them, "I know where there is some honey. Follow me, and I will show you." Now if the latter was the case, when we trapped the marked bee, the others would not be able to find the honey, because they could not follow her. But, on the other hand, if the marked bee had told her friends how to find the honey, and had described to them exactly where it was, it would not matter to these other bees whether she was with them or not. Mr. Maeterlinck's result of this ingenious experiment left the question almost as undecided as before. He tried it twenty times, but only one strange bee found the honey, which was placed in his study in the house. He asks, "Was this mere chance, or had she followed instructions received?" I have tried the same experiment a large number of times, for it interests me very much. I am bound to say that there appears to be some ground for believing that the marked bees do actually give instructions to the others, for in my case the honey was placed in a spot which was quite out of the way of the voyages of the bees, and yet on several occasions friends of the marked bees found it; and though the honey might be left in exactly the same position for a week or more before the experiment was tried, yet not a single bee ever came to it.

CHAPTER XL
BEE FLOWERS

UNTIL quite recent times it was not thought that the bees' visits to flowers were for any other purpose than to gather food for themselves. It is now known, however, that their visits are really necessary to the flowers, and it is thought that flowers secrete nectar to attract them. Some kinds of flowers contain more nectar than others, and it is not always the largest which have the most. Small flowers are quite as interesting to study, if not more so, than large ones, and there is a great deal yet to be learned about even the tiniest flower. A primrose or a snowdrop possesses wonders which even the greatest scientists of the day cannot completely fathom. Lord Tennyson knew this when he wrote these beautiful lines:—

"Flower in the crannied wall,

I pluck you out of the crannies;

Hold you here, root and all, in my hand,

Little flower; but if I could understand

What you are, root and all, and all in all,

I should know what God and man is."

PLATE XXXV

Delphinium

Nature has so arranged things that all plants do not flower at the same time. Not only does this give us flowers nearly all the year round, but it allows the bees to work many months to gather in the stores for winter. Have you noticed that as soon as one kind of flower is over its place is taken by something else? Even though this arrangement does exist, it would be of but little value to the bees, unless the flowers were "honey" flowers—that is to say, the sort which secrete good supplies of nectar. Yet the bee-keeper knows that besides the ordinary flowers, those kinds which are useful to bees also follow one another from early spring to late autumn. There is thus a sort of calendar of honey flowers all the year round.

The bees will wake from their winter sleep as soon as the fine days of spring come, and it is then that the crocus is in flower. This flower is rich in pollen, which the bees commence to carry into the hive. In March there will be the daffodil and several other wild flowers, among which we may mention the dandelion and colts-foot. In April the blackthorn and palm will appear, whilst in May there will be a large number of wild flowers ready, including the broom, hawthorn, and foxglove. But June is the great bee month, for the fruit trees in the orchards are covered with blossom, and the clover makes the fields look white. Down in the south of England, too, there is the sainfoin, a flower which gives a large amount of nectar. In July the heather attracts those bees who are near the moors, while bramble flowers cover the hedges. In August there is still the heather, but the flowers begin to go, and the bees feel that winter is drawing near, and it is now that they make preparations for their long sleep. The last flower of the year is generally the ivy, which may be seen about October. This flower gives a little nectar, but, as the days are now cold and wet, the bees seldom leave the hives to gather it.

These are but a few of the best-known flowers, for there are hundreds of other kinds, and it would be interesting for you to make a calendar of your own. The two flowers from which the most nectar is obtained, are the white clover and the heather. Some flowers are of no use to the bee, although they store large quantities of nectar, for it is so placed that the bee cannot get to it, such as the red clover.

We have seen that bees can distinguish between colours, and it is even supposed that they have favourite colours, and that they prefer blue to any other. If you are able to watch a flower called delphinium, or larkspur (Plate XXXV.), which is light blue, and grows in parks and gardens, you will be surprised to notice what a number of bees it attracts, even though there may be many other kinds of flowers around.

PLATE XXXVI

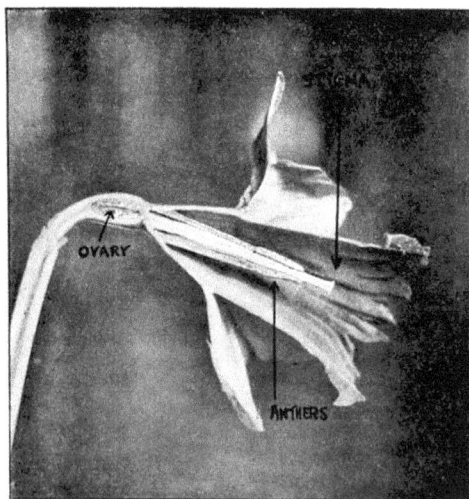

From a photograph by] [E. Hawks

Sectional View of Daffodil

CHAPTER XLI
POLLEN

WHILST it is true that plants cannot speak or walk about, yet they live a separate life of their own. They breathe and sleep, feed and digest just as animals do, but in a different manner. In order that we may understand more about this and the use that bees are to them, we must first learn a little about the construction of the flowers themselves. Let us choose a daffodil about which to speak, for it is both interesting and easily obtainable.

You will know that it is made up of "flower leaves," and that there is no calyx like that of a primrose, for instance. The corolla is a deep yellow tube, and to it the flower leaves are joined. If now we cut the flower in half, we find that there is a long rod, called the style, at the end of which is a kind of sticky knob, called the stigma; this you will see on Plate XXXVI. There are six smaller rods grouped round the style, and these are called the stamens. They are thickened at the end near the stigma, and the thickenings are called the anthers. The anthers are the pollen-bearing parts of the flower, and though their position often varies, you will find both anthers and stigma in nearly every kind of flower. Below the corolla of the daffodil is the ovary, and this is where the seeds are formed. If we look in the ovary of our daffodil, we shall see several tiny round objects of a transparent nature. These are called the ovules, and in time they may become seeds. There is a remarkable difference between an ovule and a seed, for if we planted one of the former, it would simply wither and decay in the ground. If, however, we set a seed, sooner or later a plant, like that from which the seed was taken, will spring up.

An ovule only becomes a seed after it has been fertilised, and this is accomplished by some pollen being placed on the stigma. The style is a kind of tube, and is connected with the ovary, and when grains of pollen fall on the stigma they send out long shoots, called pollen tubes. These pollen tubes grow down the style till they reach the ovary. Each pollen tube then finds an ovule, forces its way in, and pours in nutrition from the pollen grain on the stigma above. The ovules then undergo certain important changes, and are turned into seeds. Pollen grains are of all sizes and shapes, but they are generally very tiny indeed. When I tell you that hundreds of grains of the kind would take up no more room than a pin-head, you will understand how very minute and wonderful are these tiny pollen tubes.

The change in the ovules, which we have just read about, is called fertilisation, and we know that this is necessary to a plant if its ovules are to be changed into seeds. We might imagine that there is no difficulty about this in the cases of flowers where there are both anthers and stigma, but it is a law of Nature that it is not desirable for flowers to be fertilised by their own pollen. Why this should be we do not know, but it certainly is an actual fact. By this I do not mean to say that flowers cannot be fertilised by their own pollen, but that they produce healthier and more numerous seeds when fertilised by pollen from another plant. Pollen from another flower of the same plant will not do, but it should be from another plant altogether. Of course the two plants must be of the same kind, for it would not do to expect the pollen of a sweet-pea to fertilise a wallflower.

Some flowers will not be fertilised at all by pollen from their own plant, and one of these is clover. Mr. Darwin, a scientist who has taught us a great deal about this subject, tried an experiment in which he fertilised twenty heads of clover by the pollen of other clover plants. They produced no less than 2290 seeds, but when another twenty heads of clover were kept from being fertilised by any but their own pollen, not a single seed was produced.

No doubt you will be wondering why a flower is not fertilised when anthers covered with pollen surround the stigma. The explanation is very simple, for the stigma has to become ripe before it can receive any pollen. In some plants the stigma is ripe before the anthers give off pollen, whilst in others all the pollen is given from off the anthers before the stigma becomes ripe. Thus we see how Nature prevents a flower from fertilising itself.

CHAPTER XLII
BEES AND FLOWERS

FROM what you have read in the previous chapter you will see that for a flower to be fertilised the pollen must come from another plant. How, then, is this effected, for plants cannot walk to one another and ask for each other's pollen? There are two ways in which Nature's law can be fulfilled. The first is by the wind, for the pollen of some flowers, such as the willow-catkin, may be blown on to the stigmas of other catkins, and thus fertilise them. The stigmas of such plants are made branched and hairy, so as to allow of their more easily catching the flying pollen as it passes.

You will easily understand that it would not do for all plants to be wind-fertilised, for the chances of pollen grains alighting on stigmas would be very remote if that were the case. By far the greater number of plants, therefore, are fertilised in the second manner, which is by insects. The bees are the most useful of all, and we now see what service they render to plants, for when a little worker dips into a flower in search of nectar, her body becomes covered with pollen. It may be that the next flower she comes to is one in which the stigma is ripe, so that the bee, as she pushes her way in, rubs her pollen-covered body against it, and thus the flower is fertilised by pollen from another plant. When a bee is nectar-gathering, you will notice that she always keeps to one kind of plant on each journey, just as the pollen gatherers do. This arrangement fits in with Nature's plan, for it is thus that pollen of the sweet-pea is carried to another sweet-pea, and not to a wallflower, and so with each kind of plant.

Many people think that the beautiful colours and scents of flowers exist only to delight man, but this is quite a wrong idea. For instance, just think of the gorgeous flowers which must grow and die in places where no human eye ever sees them. The real state of affairs is that man uses the flowers which already exist, and even if all men were to die, flowers would still continue to blossom.

The more we study flowers, the more clearly does it become evident that their rich colours, beautiful perfumes, and sweet nectar are really baits to entice insects to visit them. More than this, even the very marks in certain flowers point to where the insect will find the nectar, just as signposts on country roads direct us to the place we wish to find. Have you noticed that flowers which have gaudy colours, like the tulip, foxglove, or hollyhock, often have no smell, whilst insignificant flowers, as the mignonette, privet, or forget-me-not, give off beautiful scents? The first kind attract insects by their colour, but the second by their fragrance. Certain flowers have their nectaries at the base of the corolla, as the geranium; others have tiny little glands, or bags, on their petals, like the buttercup.

You will know that flowers open and close at different hours—in fact it is almost possible to tell the time by watching them. The little daisy is so called, for it is the "day's eye," and it closes at sunset; but the evening primrose is only just waking when the daisy is going to sleep. Who does not know that honeysuckle gives off its sweet fragrance in the evening-time? The reason for these facts is this. The daisy is open during the daytime, because it is visited and fertilised by insects who come only during the hours of daylight. The evening primrose is fertilised by moths which fly in the twilight and evening, and so it has no need to be awake by day. We can easily see, too, that the tube-like flower of the honeysuckle is far too long for the tongue of the little bee to reach its nectar, and the corolla is so narrow that she cannot creep down it. So the honeysuckle relies for fertilisation on moths, who have far longer tongues than bees, and it emits the lovely smell at evening-time to attract them.

PLATE XXXVII

Nasturtiums

CHAPTER XLIII
HOW FLOWERS PROTECT THEIR NECTAR

A WHOLE volume could be written on the marvellous contrivances of flowers, but we must be content to describe a few. It is a wonderful subject, and one which you yourselves will be able to study quite easily.

Have you ever wondered why cup-shaped flowers—the harebell, the snowdrop, and many others—droop their heads? It is because they would become filled with rain or dew if they did not do so, and thus their nectar would be spoiled, and insects would no longer visit them. For the same reason daisies will close their petals when dark clouds come up, and will remain closed until the sun shines again. Have you ever seen a flower of the white dead nettle? It actually protects its nectaries with one of its petals, which overhangs the others, and acts like a little umbrella.

The ordinary nasturtiums (Plate XXXVII.) have the edge of the three lower petals cut into fine strips. These keep the rain from the nectar, which is situated at the end of the long spur. You will notice that hive bees are not often seen on nasturtiums, for their tongues are not long enough to reach the nectar; so these flowers depend more on humble-bees for fertilisation. The nasturtium is a flower which illustrates very well what was said about "honey-guides" just now, for all the lines on the petals point to where the nectar is to be found.

Some flowers have to protect their honey from certain insects, who wish to take it without fertilising the flower in return. Ants, for instance, are very fond of honey; and, as you can easily imagine, they are so small that they can creep right down to the nectaries without dusting themselves with pollen, or fertilising the flower. So certain flowers—like the primrose—have their stalks covered with multitudes of tiny hairs. These serve as a barricade to the ant, and prevent it from climbing to the flower above. The cross-leaved heather has its stalk and calyx covered with sticky hairs, so that not only are the little thieves prevented from getting to the flower, but they are actually held prisoners as well.

CHAPTER XLIV
HOW FLOWERS ARE FERTILISED

WE have now seen something of the contrivances of flowers to aid in their fertilisation, and in this chapter we shall consider the ingenious arrangement some flowers possess to assist their fertilisation.

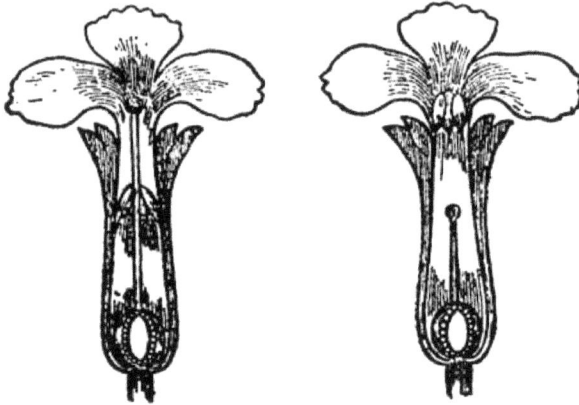

(*a*)(*b*)

Let us first look at the primrose. Have you ever noticed that there are two kinds of primrose flowers? From the outside perhaps they look very similar, but if you look closely, or better still, cut them open, you will find where they differ. Let us look at these sketches and we shall see that the one kind (*a*) has a long style, which reaches nearly to the top of the corolla. The other kind (*b*) has quite a short style, so that instead of the stigma, or knob, being at the top of the corolla, it is really half-way down. We notice, too, that the anthers, or pollen bags, in the first kind (*a*) are placed half-way down the corolla, and in the other flower (*b*) they are at the top. We might think that Nature had made some mistake here, for it seems that if the pollen bags belonging to flower (*a*) were placed in flower (*b*), or *vice versa*, things would be more natural.

Let us suppose that a bee visits flower (*a*) and dips her tongue down the corolla to collect the nectar. Half-way down the flower the tongue has to pass the pollen bags, and in doing so gets dusted over with pollen grains. The bee, having collected the nectar, flies to another plant, which we will suppose bears flowers of the other kind. She dips down her tongue, which touches the stigma just at the place where it had been covered with pollen by the first flower. By this means, therefore, the flower (*b*) is fertilised. But, you will ask, what about flower (*a*)? While the fertilisation of flower (*b*) has been going on, the pollen bags of (*b*) at the top of the corolla have dusted the root of the bee's tongue, so that when she goes to a flower of the (*a*) type, the pollen dust at the root of her tongue touches the stigma, and the flower is thus fertilised.

What a wonderful arrangement this is, for you will see that it is almost impossible for the flowers of one primrose plant to fertilise each other; the pollen must come from the flowers of a different plant.

Some flowers, if not fertilised by insects, have the power to fertilise themselves, and to this class belongs the sweet-pea (Plate XXXVIII.). This flower belongs to the *papilionaceous* (butterfly) tribe, and when a bee alights on the flower its weight presses down the underpart. While the bee is taking the nectar, the pollen bags rise and touch her on the underside of the thorax. Then she goes on to another flower whose stigma is ripe. This time the stigma rises and touches the same part of the bee's body, and in this manner the flower is fertilised.

PLATE XXXVIII

Sweet Pea

Some plants have wonderful arrangements for transferring their pollen to other flowers, some of which are so peculiar and clever that we might think they had been designed by some crafty scientist. One of these is called the salvia, and it belongs to the same family as the dead nettle. The anthers are mounted like a see-saw, and when the bee makes its way into the flower it pushes one end of the see-saw up. This causes the other end, on which the pollen bags are situated, to come down thump on to the bee's back. The pollen is thus scattered there, and the bee also receives what may be called a pat on the back! As the salvia flower grows old its pollen bags shrivel up, but at this time the stigma is ripe. It grows longer and longer, and bends

over till it is like a letter J turned upside down: ʃ After a bee has visited some young flowers and had her back dusted with pollen, she will, without doubt, visit some of the older ones too, and it is quite easy to understand that when she enters these she rubs her back against the overhanging stigma, and the pollen adheres to it.

Another interesting plant is the violet, the nectar of which is stored at the end of the long spur, which you will have noticed. The pollen bags fit closely round the stigma, and so when pollen drops from them it does not fall out of the flower, for its passage is blocked by the tight-fitting pollen bags. When the bee comes, she has to push her tongue right up the spur, and in doing this she forces it past the pollen bags. This causes the pollen to fall out on to her head, and so it is carried to the next flower.

CHAPTER XLV
CONCLUSION

ALTHOUGH very much more could be written on this interesting subject, yet there is a limit to all things, and we come now to the end of this little book.

If you did not know or care much about bees when you began Chapter I., I hope that what you have read will help you to understand something about these wonderful insects. The study of parts of their bodies, or Anatomy, as it is called, teaches us a great deal, and helps us to understand all the more clearly how they perform the duties of the hive, and how they collect their food.

Although the wonders of the hive, the combs, their building and design, the different workers and their duties, are marvellous, yet the ways of the bees themselves are far more wonderful, and we cannot fully understand them. It is not known at the present time whether the bees are able to think and reason, or whether they simply do these things by instinct. This alone is a great subject, and one on which there have been endless discussions among the cleverest scientists in the world, and yet we get no nearer the truth.

If you are not able to study the habits of the bees in the hive, there is nothing to prevent you from watching them when they are at work in the garden or hedgerow. It is always very pleasant to hear the happy song of the foragers on a summer afternoon as they flit from flower to flower on their task.

The study of flowers, or Botany, is most interesting, especially when considered in relation to insects. It was not till comparatively recent years that it was found they were connected; but one day a young German botanist, called Sprengel, happened to notice some tiny hairs growing in the centre of a wood-geranium. He determined to find out what purpose these hairs served, and ultimately proved that they protected the nectar of the little flower from the rain. From this apparently trivial discovery it was found that most plants were fertilised by insects. It seems almost as though Nature had intended flowers and insects to fit in with each other, and it is very wonderful to think of this when we remember that they belong to two different kingdoms. A great deal has yet to be learned about bees and flowers, for there are all sorts of curious devices in flowers which we do not yet understand. It is important to remember that the bees do not know that they are fertilising the flowers, for they only think of collecting nectar, and carry the pollen from one plant to another quite accidentally.

Always remember that a bee will not sting you unless it is annoyed, or unless you hurt it. If it does sting you for this reason, do not kill it, for it is only doing what it has a right to do, although it may be a painful right! I knew some boys who used to spend Saturday afternoons seeing who could kill the most bees. One day they ran to me and told me that they had actually killed 172 bees between them. Of course I told them how cruel I thought they were, but they had never thought of it in this way, and after I had shown them one of my hives and explained a few of the wonders of the bee-city, they said how sorry they were, and you may be sure they have never killed a bee since.

THE END

www.ingramcontent.com/pod-product-compliance
Ingram Content Group UK Ltd.
Pitfield, Milton Keynes, MK11 3LW, UK
UKHW031835270325
456796UK00003B/418